LOGGING

Glenn Decker topping a tree in the Alsea, Oregon area about 1940.
- Benton County Historical Society

An Oregon Logging Pioneer;

George Shroyer's Life, Work, and Humor

Biography

Joyce M. Hall
With: Lee Wood
Bert Webber, Special Editor

Reflected Images Publishers

Printed in Oregon, U.S.A.

Published by
Reflected Images Publishers
Division of
WEBB RESEARCH GROUP PUBLISHERS
P. O. Box 314
Medford, OR 97501

FRONT COVER PHOTOGRAPH
One of Alsea Area's Largest Trees.
Left-to-right:
Dan Winney, Le Roy Russell, Jack Bowen, George Shroyer.
Old-growth Douglas Fir 10½-feet diameter.
—Shroyer collection

Library of Congress Cataloging in Publication Data

Hall, Joyce M. 1931–
An Oregon logging pioneer : George Shroyer's life, work, and humor / Joyce M. Hall with Lee Wood : Bert Webber, special editor.
p. cm.
Includes bibliographical and references and index.
ISBN 1-878815-11-3 (hardcover). — ISBN 1-878815-12-1 (trade pbk.)
1. Shroyer, George Edward, 1907- . 2. Loggers—Oregon—Biography. 3. Lumber trade—Oregon. 4. Oregon—Biography. I. Wood, Lee. 1941- . II. Webber, Bert. III. Title.
CT275.S4917H35 1998
338.1'7498'092–dc21 98-22697
[B] CIP

SPECIAL FOREWORD

By: Mayme Fields Ohnemus
A Special Friend

My first recollection of George was when I was about three years old. My father worked as a mechanic for him and my mother had taken all five of us kids with her to take dad his lunch. Those were the days before car seats, and we were all bouncing around; I was the youngest. George walked up to the car and acknowledged us kids. It was love at first sight for me when I saw this balding man with his blue, twinkly eyes and kind smile.

My mom helped George's wife Helen in the yard and in her garden so I saw George often. My brothers, Laverne, Leonard, and Raymond helped mow the lawns, and work in the garden too, later they did some haying, and irrigating for George; after George's youngest son, Rex, was about three, my older sister Janet looked after him during the summers.

When I was eight years old, my life took an exciting turn. That was the summer the Shroyer's invited me to live with them in their beautiful castle like home. I had my own room and I remember listening to the birds chirping outside my windows. I helped take care of Rex each summer after that until I graduated from high school.

George took Rex and me with him whenever he could, he took us with him when he was cruising timber and fixing equipment. One day when George stopped at the filling station he allowed us a special treat—it was soda pop in long necked bottles. When we got back in the truck, George was watching me out of the corner of his eye and every time I would lift the bottle to my lips to drink; George would hit a bump, dodge a pothole, or step on the gas. Soda pop spilled all over me. I couldn't get a drop of that pop. I finally had to wait until we came to a complete stop before I could quickly drink the rest. I was surprising that George could keep such a straight face the whole time.

George would take Rex and me on the cat with him while he was building roads and landings in the woods. I loved to feel the power of the cat as it moved everything in its path while climbing-up and over logs. I always felt safe when Rex and I were sitting in the double seat, next to George.

George talked Rex and me into playing a trick on Cleo Duckworth. He told us to put some Tabasco Sauce in Cleo's chewing tobacco. George figured we knew about that stuff, he thought we'd only put a few drops in it, but Rex and I put a whole capful of that in Cleo's tobacco. After we did that, we got right up there on the cat with George. We were glued to him for the rest of the day;

figured we were going to need a little protection from Cleo. Cleo sure did spit and sputtered when he got into that; he drank gallons of water that day. I noticed that although George continued to work the cat, he kept Cleo in sight, I think George got quite a kick out of that.

Two of George's choker setters were always teasing us, so Rex and I decided to get even with them. George suggested we strip the end of a wire, and run it to a spark plug in their car so they would get a little shock when they started it. From the spark plug, we ran the wire under the floor mat, and up to the seat covering it with the bare wire. Toward the end of the day, when the men came in from work, they waved at Rex and me – we were setting on the fence and didn't want to miss anything. The guys threw their metal lunch boxes in the car, climbed in, and as they started the engine sparks flew and so did the men. They bailed out of the car and started for us kids. We barely managed to outrun them. We didn't go around them for months after that.

I pestered George all the time, I loved being around him so, but he had ways of backing me-off. One of those times was when he was fixing an old gas washing machine up there on the Idaho Ranch. George said, "I'm having a hard time telling if I'm getting any spark, could you hold on to these wires for me and let me know if you get a spark?" I was trying to be helpful so I took the wires, and he stepped on the pedal; an electric current traveled between those wires and through my body, I screamed and dropped the wires. George laughed and said, "Good, I see we have a spark."

The best summer I ever had was the year George, Helen, Georgie, Rex and I spent the entire time at the Idaho Ranch. When people would see me, they would ask George who he had with him—he'd look at me with those wonderful, twinkling eyes, and grinning, he'd say; "Oh, that's my girl." My heart felt as it would burst with pride.

After living in Alaska for thirty years we've kept in touch by visits, phone calls, and letters. I'm still "George's girl." <>

TABLE OF CONTENTS

High Climber Fred Gee poses atop a spar tree.
See text page 187.—Benton County Historical Society

George Shroyer with spread arms stands against butt of giant spruce on Eckman's Slough near Waldport, Oregon in 1942.
—Benton County Historical Society

AN OREGON LOGGING PIONEER
This is the Story of George Shroyer's Life, Work, and Humor

PREFACE AND ACKNOWLEDGMENTS

I first met George Shroyer and his wife Helen in May 1995, after Joyce Hall, their daughter, said George wanted to retain me to help him write a book. To my surprise, I found both George and Helen to be the youngest eighty-eight-year olds I'd ever had the pleasure of meeting. George still had a twinkle in his eye and a boyish grin; and Helen smiled deeply and easily during that first meeting. As they began telling me stories about their lives, I found them uncommonly etched with a deep and vigorous love of life. Theirs were not wasteful disparaging lives, but lives filled with determination and direction. We were sitting in a warm kitchen with our chairs around George's position near the heater and his own desk. This was in the kitchen where he kept special things for his grandchildren.

George said he'd wanted to write a book for the last twenty years, I assured him this would be possible. Since that meeting Joyce and I have worked together for over two years helping George document his life.

George Shroyer was an early logger in Oregon. As a young man, in the early twenties, he worked in mills that were using horse logging, and continued working while logging practices developed through an unprecedented mechanization of the industry. His life's work is told through humorous incidents, interspersed with detailed descriptions of family life and of logging practices. By recording his story, George will be passing on the essence of the camaraderie enjoyed by the logging pioneers in Oregon.

He wishes to thank all the "Old-time Loggers" who have contributed information for his book and have shared with us many stories they laughingly recalled, even when the joke was on them. He also thanks his sister Marie Kirkham, Hazel Lowther, Sam St. Clair, John Eagleson, Bud Carey, Robert Walter, Mayme Ohnemus, Gene Byler, Henry Arbuckle and his son, George E. Shroyer, Jr., his son-in-law Charles Hall, and the Benton County Historical Society and others for contributing photographs A special thank you to Teresa Schulte for

her charcoal drawings and to Don Ferguson of Shawdowsmith Photographics of Corvallis.

Mr. Shroyer especially thanks his daughter Joyce Hall for many months of detailed preparations for this book, and his son George Shroyer, Jr., and his son-in-law Charles Hall, for their assistance with technical editing. He also thanks Helen, his wife, for pushing him to write the book and for taking an active interest throughout the project and for her many contributions.

Joyce's daughterly love exemplifies itself through her dedication to clarifying and organizing his life's story. She took full responsibility for editing her father's manner of speaking throughout the book thereby creating his voice. The sequence within George's stories is all mine, but the words are all his; whereas the background information and chronicling of their times is mostly mine, a crystallization of many hours of conversation.

George's family played an active role in his life in his early years as it does now. Surrounding George and Helen are their children and grand children; the home they lovingly call "The Castle" forms the hub for their lives as they continue their life's challenges.

The first chapters, are dedicated to future generations of Shroyers. We have included a clear and concise documentation of George's Parents and Grandparents. The Shroyer Families cooperative efforts have resulted in uncovering several documents one-hundred-years or older; making it possible to authenticate the family history. We are basing our information on the recollection of stories by the oldest living family members. Additionally, details from old cards, letters and from backs of old photos have been sources of information.

The family history, being covered in the first six chapters of the book, is followed by George's earliest logging experiences. The depression and its effect on logging and family life ends, for purposes of the book in 1938. The impact of World War II on logging and family life is in chapters thirteen and fourteen. Beginning in 1945, the story turns to George's own business ventures and tells how he struggled to create his own logging business in Oregon. Then follows information on additional business ventures as a ranch in Idaho, the Detroit Dam clearing project, and a sawmill experience in Canada – the latter proved to be George's financial undoing. The final chapters cover modernization of logging practices, travel, and something of George's life today.

Lee Wood
Corvallis, Oregon

CHAPTER 1.

BEFORE THE TURN OF THE TWENTIETH CENTURY

GEORGE EDWARD SHROYER'S
FATHER—OLIVER JAMES SHROYER
born - Oct. 4, 1866—died - Dec. 20, 1961

George Shroyer's father, Oliver Shroyer, influenced his life by providing him with two basic strengths. First, he taught him to work hard; furthermore, he taught him to take responsibilities seriously. George was the oldest son and Marie, his youngest sister said, "When Dad wasn't home George drove the horses, looked after the chores and farm."

Oliver's own father died when he was only thirteen, leaving him to work the family farm, which lay about twenty-five miles north of Warsaw, Indiana. Oliver soon learned the value of hard work and accepted responsibility early in life, he also valued education. Oliver began his study of medicine in 1888; he was one of about five-hundred students attending Valparaiso University near Chicago.

> **Seems to me, that Dad was about third in his class. He might not have been smarter, but I think he studied harder, and had a real good memory. Dad being third in his class was about the same, as being first because; he found out that if you were the smartest, people didn't like you. So he would miss two or three questions on purpose. Now I wouldn't do that I don't think. I know people are that way, Helen Hunter, the young woman I married, and Hazel Ellis were highest in their class. The girls were real jealous of them. I never had any trouble that way, I missed enough everybody should have been real friendly. Everybody should have liked me, because I never got perfect grades, you know.**

Oliver found that Medical school was not all memorization and dry academics, when there were research projects, interest peaked.

> **I remember Dad telling about an experiment they did one time. About a dozen of them got together and decided that they would try something the professor had told them about. They were going to see if they could, by suggesting that a guy was sick, make him sick. They decided to try it on the old janitor everybody called "Dad". When the janitor first came in that morning, one of the students met him at the door and said; "Dad, you're sure looking fine today, how are you feeling?" He said; "Oh, just fine." So, the guy wrote it down, so they could have a record of it. When another student saw him a little later that morning, he said; "Dad, you look kinda peaked to me." He said; "Oh no, I feel fine." Later that morning, after they'd told him how**

Valparaiso University where Oliver studied medicine for two years.

sick he looked, and wanted to help him up and down the stairs, he decided that he was sick. He asked if he could go home, and they sent him home.

While Dad was studying to be a doctor, he got a job in a doctor's office working part-time to help pay his way. When a patient would come in to see the doctor, he always had Dad go into this little closet where they kept their medicine and stuff, it had a curtain over it. He could hear everything that went on and could learn a lot that way. Dad said the doctors had some medicines that worked real good, things like digitalis and morphine; but some of the medicines were made of nothin' but chalk. The doctors made these pills in different colors. They gave a different color for each sickness; as a pink one for infections and a yellow one for stomachs. They gave Smith and Jones the same pill if they had rheumatism; if they didn't, the neighbors would think the doctors didn't know anything. If a flu would go around town; they would give them all a blue pill.

Well, they were giving them that chalk and they would charge them for it. The first call was two-bits, then if they had to keep coming in, it was only a one-time charge of a dollar. A doctor was paid in chickens, squash, cabbage or money. Dad said there was so much quackery about it, that he got so he didn't like it; after about two years he decided to become a teacher. So Dad got a teaching certificate and a business degree instead of a diploma in medicine.

DIPLOMA

NORTHERN INDIANA BUSINESS INSTITUTE.

In Omnia Paratus

VALPARAISO, INDIANA.

To whom this may come greeting:

Be it known that Oliver J. Shroyer this day has completed, at this Institution, the Business Course which embraces the following branches to wit: Bookkeeping both Single and Double Entry Commission and Shipping, Jobbing, Partnership, Banking, Rail Roading, Steam Boating, Commercial Correspondence, Commercial Arithmetic, Penmanship, English Grammar, Commercial Law and Debating.

In testimony whereof we have hereunto subscribed our names this 2nd Day of June MDCCXCII

H. B. Brown President.

C. W. Benton Science of Accounts.

H. B. Brown English Grammar.

J. B. Showalter Commercial Arithmetic.

Debating.

Penmanship.

Oliver's diploma from Northern Indiana Business Institute is dated "2nd June MDCCXCII"

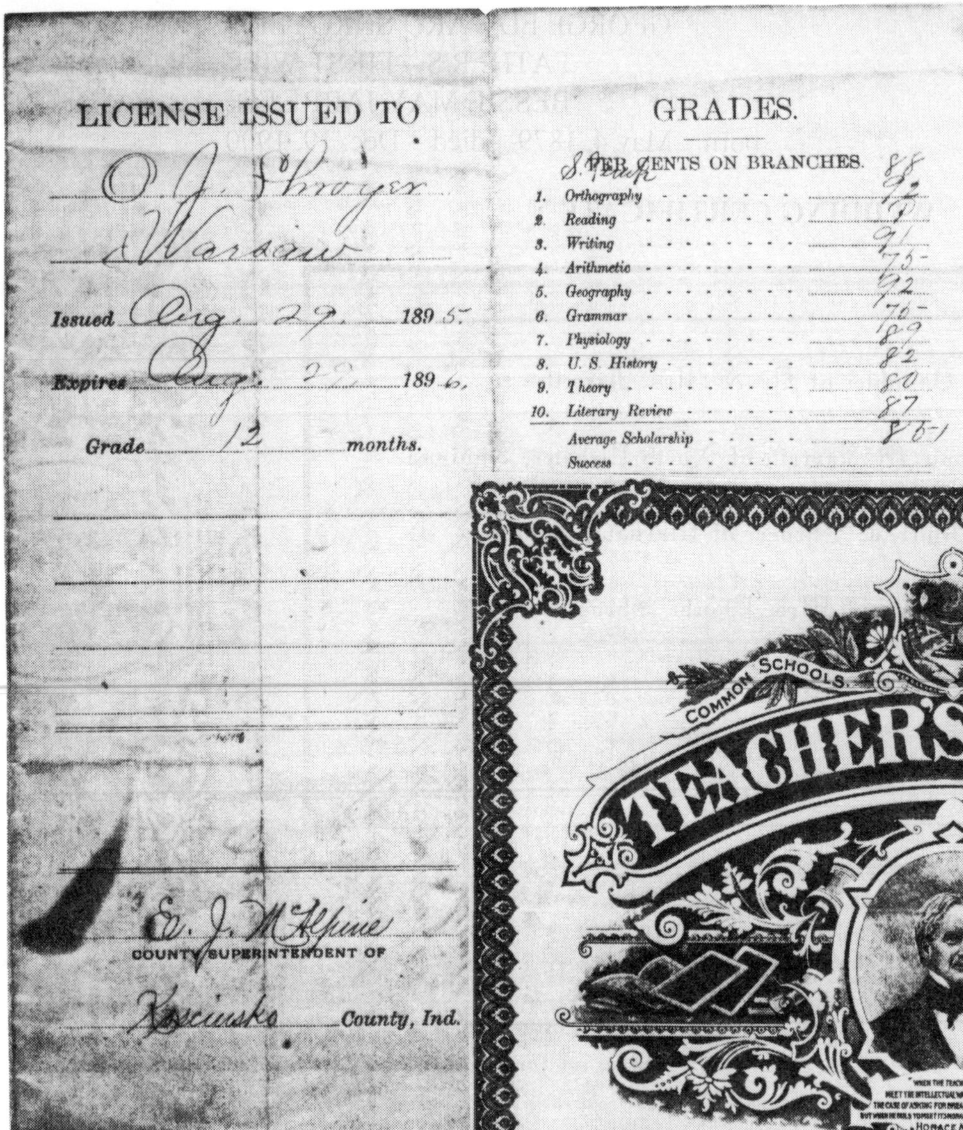

LICENSE ISSUED TO

O. J. Shroyer

Warsaw

Issued Aug. 29. 1895.

Expires Aug. 29 1896.

Grade 12 months.

E. J. McAlpine
COUNTY SUPERINTENDENT OF

Kosciusko County, Ind.

GRADES.

PER CENTS ON BRANCHES.

1.	Orthography	88
2.	Reading	92
3.	Writing	91
4.	Arithmetic	75
5.	Geography	92
6.	Grammar	70
7.	Physiology	89
8.	U. S. History	82
9.	Theory	80
10.	Literary Review	87
	Average Scholarship	88.1
	Success	

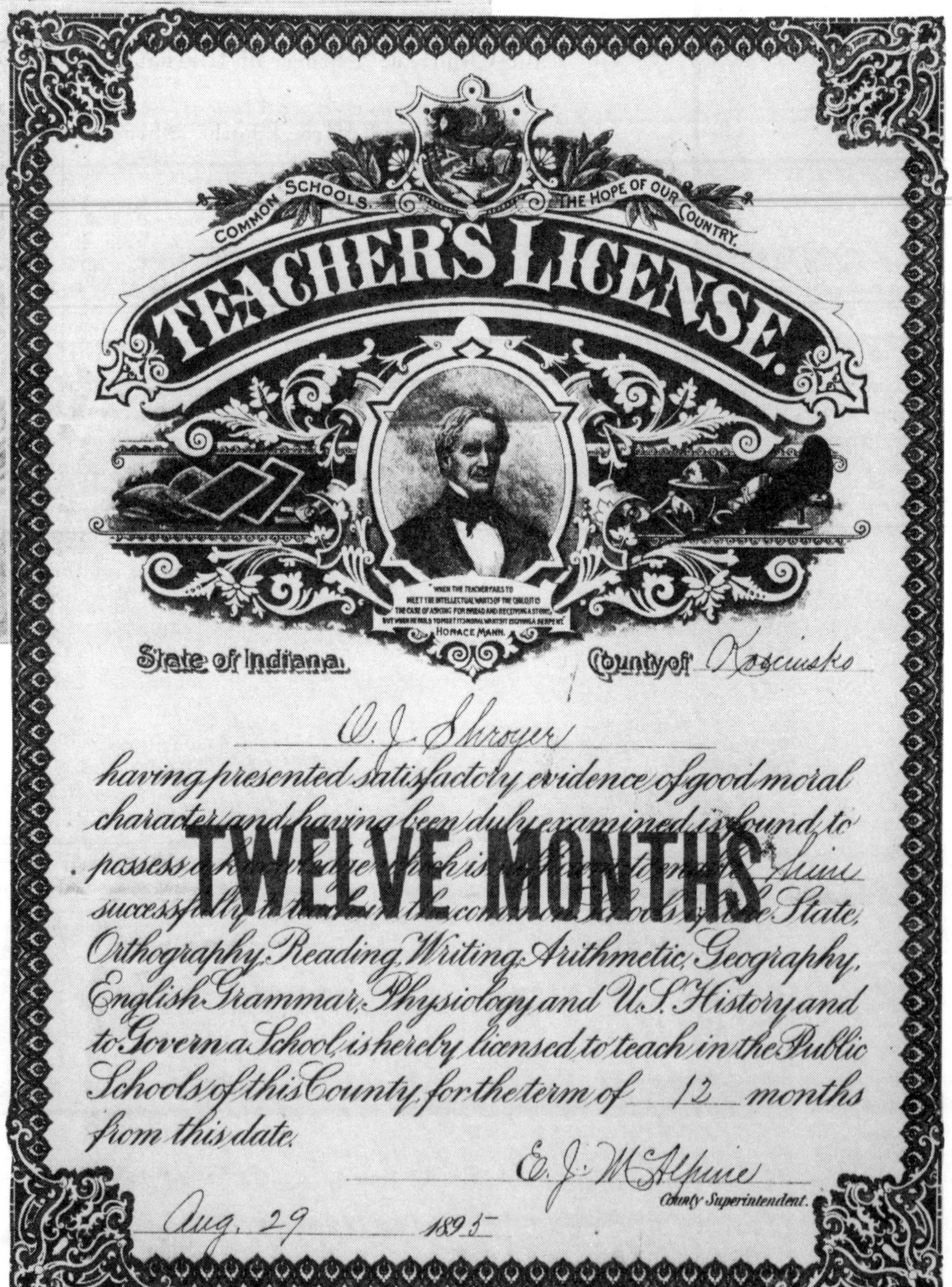

State of Indiana, County of Kosciusko

O. J. Shroyer

having presented satisfactory evidence of good moral character and having been duly examined is found to possess a knowledge which is sufficient to teach successfully in the common schools of the State, Orthography, Reading, Writing, Arithmetic, Geography, English Grammar, Physiology and U. S. History and to Govern a School, is hereby licensed to teach in the Public Schools of this County, for the term of 12 months from this date.

TWELVE MONTHS

E. J. McAlpine
County Superintendent.

Aug. 29 1895

Oliver Shroyer's Teacher's License for State of Indiana dated "Aug. 29, 1895" and his final grades.

GEORGE EDWARD SHROYER'S
FATHER'S—FIRST WIFE
BESSIE MAY JARRETTE
born – May 4, 1879—died – Dec. 19, 1900

WEDDING CERTIFICATE

Daughter of Dr. & Mrs. Jarrette

Miss Bessie M. Jarrette of North Webster, Indiana

and Oliver J. Shroyer of Warsaw, Indiana,

Son of Mrs. Phoebe Shroyer

Were United in Marriage

This 27th day of December 1896.

OLIVER AND BESSIE SHROYER'S HOME IN INDIANA

Oliver built a home for his first wife, Bessie, in 1898. Bessie bore him a son, Ernest Eugene, on November 25, 1897. This child lived only five weeks. He died on January 2, 1898. Their daughter Eva Shroyer was born the following year on May 7, 1899. Bessie died the following year of galloping consumption, also known as tuberculosis, on December 19, 1900.

Mrs. Phoebe Shroyer, George's Grandmother.

GEORGE EDWARD SHROYER'S
PATERNAL GRANDPARENTS

MR. GEORGE SHROYER
born – June 6, 1808—died – June 11, 1880

MRS. PHOEBE COUNTS BENNETT
born – November 5, 1832—died – June 12, 1909

George Edward Shroyer's grandfather, George Shroyer, was born in Rockingham County, Virginia on June 6, 1808. As a young boy, he moved with his family to Ohio. He was a carpenter during the early part of his life. In Ohio he married Sarah Wolfe, they had one daughter and six sons. Later, he settled near Benton, Indiana as a farmer. His first wife died in 1861 and he married a Civil War widow, Mrs. Phoebe Counts Bennett, who was born November 5, 1832. She was the daughter of James and Sallie Counts, who later settled in Montgomery County, Ohio in 1839. George Shroyer and Phoebe Counts had two children by this marriage, Clara A. and Oliver J. Shroyer. George Shroyer died June 11, 1880 in Benton, Indiana. Phoebe Shroyer moved to Alhambra, California and died there June 12, 1909. George Edward Shroyer believes he can remember setting on his grandmother's lap in her home in Alhambra, California.

CHAPTER 2

STARTING OUT IN CALIFORNIA

GEORGE EDWARD SHROYER'S
FATHER—OLIVER JAMES SHROYER
born – Oct. 4, 1866—died – Dec. 20, 1961

Oliver, having lost Bessie, was concerned for his young daughter. The doctors felt she may have contracted tuberculosis and were advising Oliver to move to a warmer climate for his daughter's health. Oliver then moved from Indiana to California in 1903, accompanied by Phoebe Shroyer, his mother, and Clara Alice Shroyer, his sister. They cared for his young daughter, Eva Shroyer. They settled on the West Coast in Alhambra, California. There he began working for Gail Borden, over-seeing the finishing work for the Borden Construction Company.

George recalls an interesting story about Gail Borden's father and how he was able to make money by patenting the processing of milk.

> **This story took place during the gold rush days when Old Man Borden, went on a ship to Alaska and on it were, mostly men, but two or three families and one was an Irish Lady with a bunch of kids. After the second or third day, most of the kids were all crying around, they wanted milk but it was all soured. This Irish Lady, her kids weren't crying. So, Old Man Borden asked her; "The other kids are all crying around, but yours are not, how is that?" She told him; "Well, my milk's not sour." He asked her how come and she said that she brought it to a boil, just as she did her fruit when she canned it. So as far as I know, he didn't make money gold mining, but when he came back, he started the Borden Milk Company. That's where it started, you may still buy canned Borden's Milk in the stores today.**
>
> **Dad always said; "Gail Borden was a real nice guy to work for and he named my brother after him." When he died, Dad still had the tools he'd used for making ornate Gingerbread Trim while he worked for Gail Borden.**

GEORGE EDWARD SHROYER'S
MOTHER -- MABEL LOUISE HATCH
born – April 3, 1878—died – Nov. 2, 1956

Oliver also taught Sunday school at the church there in California, one of the couples he'd met was Tracy and Effie Hatch. Effie had become ill with consumption while in Illinois, so the young couple had moved to California so she could regain her health. Tracy's sister, Mabel Hatch had come from Chicago to care for Effie and their two young daughters, Hazel and Ester, until Effie could recuperate from her illness.

They introduced Oliver to Mabel Hatch, a recent graduate of the school of nursing. She was a good match for the well-educated man Oliver Shroyer. Oliver Shroyer and Mabel Hatch married on April 11, 1905, and made their home in Alhambra, California.

Mr. Oliver J. Shroyer

Miss Mabel L. Hatch

OLIVER J. SHROYER AND MISS MABEL L. HATCH

WEDDING ANNOUNCEMENT

Mr. Oliver J. Shroyer

Miss Mabel L. Hatch

Announce Their Marriage

On Tuesday the Eleventh of April,

One Thousand Nine Hundred and Five.

Oliver and Mabel lived in their comfortable home, built and owned by Mr. Oliver Shroyer. They soon began a family and by 1908 had three young children and Mr. Shroyer's child Eva, from his first marriage. Oliver and

Mabel Shroyer's children were; Lucy, born on March 28, 1906; George Edward, on May 21, 1907; and their youngest son, Gail Raymond, on November 2, 1908. Marie, their youngest child, was born on April 4, 1916 in Oregon. She recalls her mother saying of Eva's mother; "Poor woman, she died so young, lived only twenty one years." Mabel always claimed Eva as her own daughter.

OLIVER AND MABEL SHROYER'S HOME

IN ALHAMBRA, CALIFORNIA

(TOP) Oliver Shroyer holding Lucy, Eva standing, and Mabel holding George.
(LEFT) Eva with her father Oliver Shroyer. Eva was 9 years old.

GEORGE EDWARD SHROYER'S
MATERNAL GRANDPARENTS

MR. EDWARD PAYSON HATCH
born April 17, 1840 -- died November 26, 1921

MISS ANNA STARR BARBOUR HATCH
born August 9, 1847 -- died February 25, 1889

Mr. Edward Payson Hatch served in the Civil War, this young soldier from Illinois received some hand knitted socks as a gift for Christmas. Miss Anna Starr Barbour had placed a label inside the socks, giving her name and address. He began writing to her. After the war he looked her up, they married and had four children. Tracy B. Hatch, born in 1870, Mabel L. Hatch, born in 1878, James Hatch who died in infancy and Ralph Raymond Hatch, born in 1882.

> **Grandpa Hatch worked in Chicago for a company that manufactured sewing machines, the type used in industry. During this time, he designed a mechanism for sewing machines for the overcast stitch in underwear and flat seams.**
>
> **Later he became a mining engineer, so he moved around a great deal. He was mixed up in mines, sometimes he had lots of money and sometimes he'd be clear broke. Once, he owned a silver mine in Nevada and he had a patent for a "Wild Rose" pattern on silverware.**
>
> **Grandpa Hatch's son, Tracy Hatch was also an inventor. It is said he had about two hundred patents. His last patent was for an automatic shell launcher for those large guns used on battleships, he invented that during World War II. The last time I visited him, he showed me his "Certificate of Recognition" signed by President Roosevelt. Tracy died shortly afterwards.He was nearly eighty years old.**

Anna Starr Hatch, Mabel's mother, was born in 1847 and died in 1889, when Mabel was only eleven years old. Tracy, her older brother entered a military academy while Mabel and Ralph went to live with their Aunt Emma Barbour Raymond, her mother's sister, who had no children. Years later, her father married Clara Cordelia Smith Hatch, her step-mother.

Mabel often visited Uncle Frank Barbour, her Aunt Em's brother. He became an investor at an early age and made millions through the Chicago Board of Trade. He and his wife had no children so when he died, he left his fortune to charity.

Mabel Louise Hatch enjoyed the close company of her cousins Mabel E. and Rose Hatch who also lived in Chicago. They spent many warm summer days together celebrating holidays as a family.

Mabel was a well bred woman of her times. She lovingly referred to her Aunt Emma as "Aunt Em." Emma made certain Mabel's education included finishing school as her mother would have wished. The Chicago finishing school included the social graces, classical literature and Latin. She also accomplished all of the school figures required for ice skating. She became engaged while quite young, but her betrothed died in the Spanish-American War in 1898. He is buried in Arlington National Cemetery.

(TOP) Oliver and his daughter Eva. He had lost his wife Bessie. (LEFT) Mabel Hatch, center, with her cousin Rose Hatch and Rose's son.

A NURSING SCHOOL GRADUATE

Mabel Louise Hatch went to nursing school and was graduated as a nurse.

She later moved from Chicago, Illinois to Alhambra, California with her brother Tracy, and his wife Effie.

Mabel drove Tracy's Cadillac while she lived with them in California.

It was during this time that she met and married Oliver Shroyer.

Tracy and Effie Hatch and their two young girls lived just around the corner from the new home Oliver built for Mabel. The girls were about Eva's age. Hazel was born in 1900 and Ester in 1902. They were six to nine years older than Lucy, George and Gail. These families were very close.

> **I'd ride in my stroller whenever we'd go to Hazel and Ester's house. We'd have to go past a bad dog that I was scared of. He'd bark and growl, and jump up on that fence, as if he'd like to get hold of us. I was sure glad when we'd get by that place.**

(LEFT) George Shroyer in his perambulator. (RIGHT) George Jr. as a toddler.

Another time, when I was in the baby buggy they were taking me for a walk, and a "mad dog" came after us. Eva grabbed that buggy, she and my cousins took me, and ran to our yard; fast as we could go.

I remember seeing that dog. Yeah! It was a big dog, it was black and white; it barked and growled and tried to bite us. I was probably two or three years old. They kept little kids in buggies a lot longer then than they do now; any time you'd go to town you'd ride in the buggy.

I remember Mama being upset about that, she thought the dog was rabid. She was a trained nurse, you know. I remember the girls being all excited too. It was a strange dog, it hadn't been around the neighborhood before. I don't know what became of him, we never saw him again. Mama and Eva talked about that for years.

I remember playing around in the yard. They had cement walks at that time, but nothing else, and the road was gravel and rocks. The walks would get so hot they would burn my feet. We kids would be playing around barefoot. Boy! You'd get back on the grass in a hurry and stay off the sidewalks in the summertime after the sun come out. We didn't wear any shoes when we were playing; took them off like the other kids. I spent my first four summers there; guess that's why I like the sun so much. I was about four-and-a-half years old when we moved to Oregon.

Now Oliver was a forward looking man, he became interested in the newly opened Oregon Territory. Homesteaders had cleared much of the area for farming. He wanted to buy some of this land, so he traveled to Oregon in October of 1910 and bought a three-hundred-sixty acre partially-cleared homestead. The Summit School hired him as school principal for the following year and Oliver Shroyer moved his family to Summit, Oregon in 1911.

CHAPTER 3

THE JUMPING OFF PLACE

A cold damp sunless day greeted the young child, George Shroyer, as he stepped from the train after the grueling trip from California. The train ride to Summit, Oregon had been round-about and slow as Summit was not on the main line.

George sat on the bench outside the station looking at the distant Douglas Fir reaching their fingers toward the sky. He did not know that Oregon would become his beloved home in which he would spend his remaining years. Nor did he know, that in his struggle to raise his family and provide jobs for others, he would play a role in Oregon's logging future. He would build numerous roads, bridges and clear an area for a dam that would provide water and power for the growing population.

George, Lucy and Gail went into the train depot where it would be warmer. George watched quietly over his younger brother Gail while wondering whether he would have any playmates here in this new place. His parents and Eva, his older half-sister removed the luggage from the train. They placed it on the wagon that would soon take them to their first home in Oregon. Oliver had arranged to have their closest neighbor, Hymes, pick them up at the train station. Hymes assisted them in loading the wagon with their belongings. The sky was beginning to darken as evening approached so Eva scurried to the post office to send a post card to her cousin, Miss Hazel Hatch in Alhambra, California to let her know they had arrived.

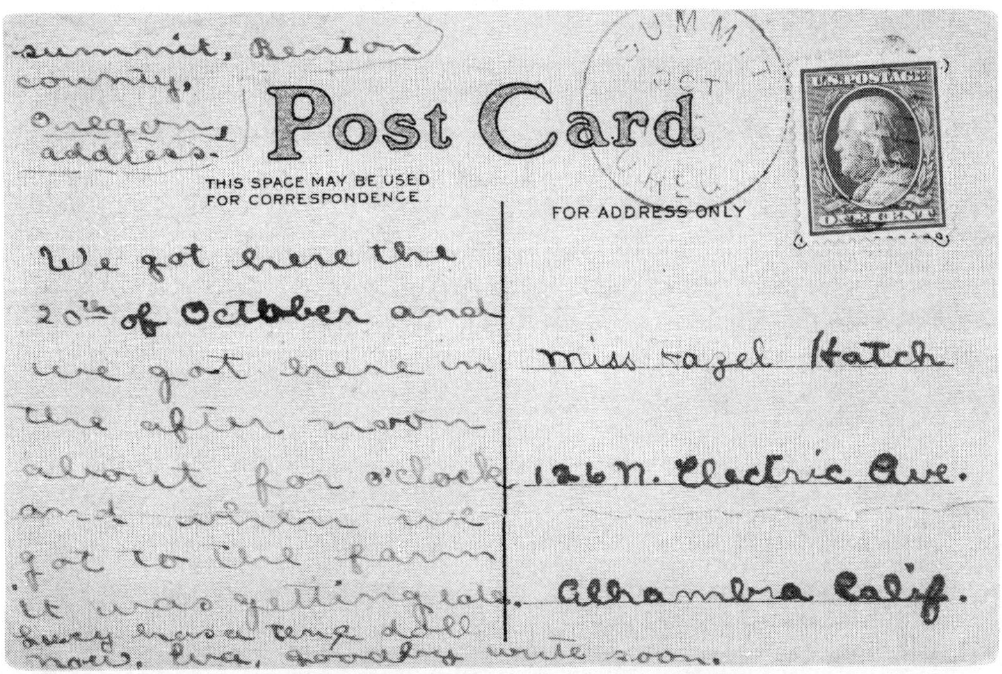

Original Post Card sent October 20, 1911

"We got here the 20th of October and we got here in the afternoon about four o'clock and when we got to the farm it was getting late. Lucy has a doll now. Good by write soon. Eva"

As the wagon carrying its passengers traveled the two-and-a-half miles to their new home, the trees along the road increased in size, there were many Douglas Fir, and as they neared their destination they saw several large Cedars. As they rounded a bend in the road, Oliver began pointing to the place they would call home for the next year-and-a-half. George could see that it was not at all like their place in Alhambra, California.

This place was a large wooded area with only a small, roughly hewn log cabin, and an addition on one end, built of lumber. The cabin set in a clearing on a knoll, this was the structure that they moved their belongings into. The log portion had no floor other than chips and dirt. The adults arranged the cabin, started a fire in the stove then created temporary beds for their first night in this new place. George wandered about the property staring in wonder at the tall trees. These trees would, in the future, supply their family with warmth for many years.

This day would mark the beginning of a new life for the Shroyer family. Their mother, a Chicago woman, was not use to this sort of life, she and Eva did not like the idea of moving to Oregon which they thought was at the end of the world. They would one-day, refer to this place as the "Jumping-off Place."

The ride to their new home was a memorable one; a day filled with wonder and anticipation. Gail would have his third birthday on November 2, 1911, just a few days away. He would be the first family member to celebrate a birthday in Oregon. Why did this man decide to bring his young family to Oregon, and why did he decide to come to Summit?

Perhaps he realized that his young children hardly knew him. Certainly, this dramatic change in Oliver's profession had been well thought out. He became the central figure in his own children's lives. He was now responsible for the development of their ethics, character and abilities. He would spend countless hours working with his children and overseeing their education. This new life would directly involve him in training them to work with, and care for their animals as well as providing the wood needed for their family's warmth. The cultivation of the garden was something Oliver actively engaged in with his children. Surely the children would benefit from this close association with this intelligent and dedicated man.

> He'd often said, 'He was afraid for us kids, that growing up in the city wouldn't be good for us.' I kinda wondered if maybe he thought the city couldn't stand us, you know.
>
> Right after we got there, Dad had a neighbor come in and buzz up several cords of wood. He had lots of other things to do that first winter. He also put in a temporary wood shed since the rain in Oregon would cause the wood to mildew over-time and, wet wood didn't burn well.

The "Jumping Off Place"
at the end of the world: Oregon
—Original charcoal by Teresa Shulte

Life would be easier later on. Oliver would establish a routine that would ensure plenty of dry wood for his family, and there would be plenty of food. They would all work at their chores and would learn the worth of their summer-time labor when the dreary-wet winters arrived.

There are only a few details that George remembers about the place his family called home those first two winters. Oliver had a great deal to accomplish; building the family a new home, preparing new fields, the land for the garden, and their new berry patch.

> **The cabin we lived in first was real small, the log part was kinda like a wood shed; you couldn't live in it. In one corner of that was a big old stove, it was mostly used for heat. I think Mama did some cooking on that stove, not much. We lived in the addition; it was just one big room built of lumber. There was an old cook stove against the far wall. It was not a good stove, but she just made do with what we had. She used this corner for her kitchen.**

We got some chickens from the neighbors, right away, that first year. Mama and Eva took care of the chickens, so we had plenty of eggs in the spring, summer and fall. When winter started, we had plenty of chicken to eat.

Those chickens use to get in under the addition that we were living in; they were always going underneath there. We had a chicken house for them; but mostly they would go under our house. That's why those darned old hawks would come after our chickens all the time. So Dad got some Guinea hens and after that, whenever those old hawks would come after our chickens, those Guinea hen's would take-off after those hawks. Right in the air they'd go—takin'-off after those darned old hawks. Those hawks better be pretty fast; cause if they weren't, one of those Guinea hens would get them. They would whoop down and kill 'em.

Later Dad started a good sized berry patch near the new house, he put a picket fence around it. Then he built a chicken house at one end of that berry patch so the chickens would keep that patch all cleaned up. He even planted some berry vines right up against the chicken house. The chickens weren't too difficult to take care of after we moved into our new house.

I remember Mama and Eva had lots to do. The old house had lots of strawberries growing in a big patch near it. Yeah, it had real good strawberries. They moved some of the strawberry plants from that patch over to our new berry patch. About half the strawberries in that old patch were white, they got bigger than the red ones. I thought the white ones wouldn't be good to eat as I thought they had to get red before you could eat them. I saw Mama and Dad eating those big white ones; so, pretty soon I tried eating some of those too. They were a real good tasting strawberry, but they were white. Later I found out they called them White Globe strawberries, some of those would be more than two inches in diameter. I remember Mama made strawberry jam that spring. Boy! That was good on freshly baked bread.

Any woman's patience would be tested by the inconveniences encountered during this first year-and-a-half. Mabel dealt cheerfully with the trials of these primitive circumstances. Caring for a large family housed in such small quarters was not an easy task, this certainly tested, even her strength of character. Mabel seldom lost her patience; she firmly believed in training her children through example. Believing the children would follow her example, not her words; made it her responsibility to be a good example of a caring wife and mother. She made do with what she had, and certainly did not complain. This ability to maintain her composure, was surely instilled in her, during her years at finishing school. Some of her earlier training no doubt stayed with her throughout her life.

Oliver would begin teaching at the beginning of the next school year. He and Eva would walk two-and-a-half miles into Summit each day. The school would pay him sixty dollars a month for six months, totaling three-hundred-sixty dollars a year. This, plus the income he earned as a carpenter each summer provided for the family's needs. Oliver sold the property and houses he'd owned in Alhambra, California to help pay for the farm.

> **Dad owed one-thousand dollars on the farm when he bought it. A year's interest on the money he'd borrowed was one-hundred dollars per year and taxes were one-hundred dollars each year. That left only one-hundred-sixty dollars for food and supplies. He usually built a house or a barn during the summer months while he was off. He'd get some money and some supplies for that, one time he got lots of apples. I don't know how much money he had, but he didn't have much.**

Since Oliver had no job that first winter; one morning he took his little box of tools and walked three miles over to Manning Underhill's. He was the owner of the only store in town and he was building a barn. Oliver told Manning that he would work until noon, if his work was not satisfactory there would be no charge and he would just take his tools and go home. If he liked his work, and wanted him to stay—just tell him.

> **When it was noon, they invited him in to eat dinner with them. Then Underhill told him they wanted him to come back, that's how he got his first job in Oregon.**

New fields would be needed to supply hay for the livestock. Oliver wanted to grow at least half the grain needed by his family but this year he would have to buy all of the wheat from his neighbors. He would also have to buy some equipment to winnow the wheat for removing the chaff. He still had the grinder his father had used in Indiana, it had the Shroyer name stamped onto the wood base. (See photographs page 122)

> **Dad thought we needed more land to grow hay on. The trees had been felled and bucked, but not hauled off, so Dad hired a guy with an outfit to clear up the land. The first year we were there; that guy pulled the logs off, and put them in piles so we could burn them. I remember going out there and watching him pulling them with a team of oxen. Next year, he had the man clear the fields down low; there were two pretty good fields in there. The field to the east of the new house seemed pretty big to me, it might have been twenty acres, there was a good orchard in there. There was another field, maybe fifteen acres, we worked clearing that every year until we had about forty acres cleared.**

Oliver needed to start the new house so the family could move into it by the second fall. The children were getting bigger and he would need to provide a better place for Mabel to care for the family. For the foundation, Oliver choose

big cedar logs, he also made cedar shingles for the roof. He ordered the rest of the lumber and siding for the new house. It would be quite large. This would be a two-story place, with three bedrooms up stairs and one downstairs. Also, they would have a living room, parlor, dining room, and kitchen.

Clearing land with a team of oxen.
—Original charcoal by Teresa Shulte

The new house Dad built us was a-half-mile to the West of the log cabin. When he wrote out the order for the lumber, he figured the amount so exactly, there was only one wheel-barrow full of wood left-over, when Dad finished building it. He ordered the lumber from the planer mill in Philomath. It was cut in a sawmill at the foot of Alsea mountain and they brought it down five miles into Philomath in a flume.

I remember Dad even had quite a bit of his furniture from Indiana to put into the new house. He also had some Indian relics that his folks left him, and some that he found. When he brought those things up on the freight train it wrecked off some bridge, and he lost most of them. Dad always hated it, that he didn't have many relics left. He lived to be ninety-five and he did his best to preserve family treasures. He still had his college-class photo, diploma, and carpenter tools for making Gingerbread Trim. (For tools see photographs on page 122)

Oliver and Mabel's new home near Summit, Oregon. George and Gail on left. George was six years old when the house was finished.

There was more open ground where the new house stood. It was higher and had a better view. There was a spring not more than seventy-five-feet away that provided the family with sparkling water year around. Oliver took care to protect the spring by putting a four-foot-diameter cemented-tile in the ground to cover it.

We were lucky! We had "runnin' water," we'd run outside and get it. Although I was only six-years old, I remember doing that. Later, Dad piped the water directly into the house to a hand pump inside.

That first year, Dad planted a big garden. It produced corn and potatoes that fall. Dad and I would go down to work on the house taking butter and salt with us. Each morning, he would start a fire to

(TOP) **Summit, Oregon in the early 1900's. The village was on the watershed between the Willamette Valley and the Yaquina River. Summit was in Benton County almost to the Lincoln County line.** (LOWER) **Summit General Store, established 1901.**

Southern Pacific railroad depot, Summit, Oregon with Manning Underhill's General Store in background.

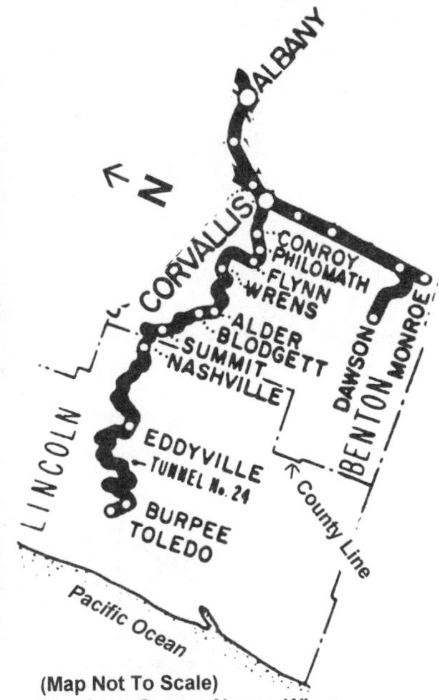

(Map Not To Scale)
Map from *Oregon Names Where They Are and How to Say Them*

> warm our hands. Later, at about eleven o'clock, Dad would put roasting ears, with their husks still on, and potatoes underneath the coals; then we'd go back to work. About noon, when we'd stop to eat, he'd just uncover the potatoes and corn, add butter and salt. Boy they were good.

Settling in their new home that second summer was an exciting event. They readied their order to Rice and Phelan, a bulk-foods company for their yearly staples in bulk. This saved money. This purchase included about ten sacks of flour, a hundred-pound sack of sugar, macaroni, beans, rice, etc. In a couple of weeks, their order arrived at the freight office. The family rode into Summit in the buggy, to pick up their supplies which would have to do until next year. They couldn't get a buggy in, or out during the winter time. Later in the spring, when the roads were better, they could pack it in on horse back.

> If the weather wasn't too bad, Dad always hitched up the wagon, and took us to Sunday School and Church. When the weather got bad, about the first week of November, it would become quite muddy, and we wouldn't go to church on Sunday very often, unless we walked. Dad and Lucy liked to go to church. Lots of times just Dad and Lucy would go even if they had to walk. Usually Mama didn't go unless we took the team in as she didn't like the trip since the trails were mostly mud. It was five miles, in and back, and there weren't any sidewalks.
>
> Winter weather settled in early, up there at Summit; snow could fall any time after the middle of November. It was just high enough to have snow instead of rain – elevation 729-feet. The Southern Pacific Railroad always stopped at Summit. This was the highest point between the Willamette Valley and the coast therefore that's where it gets its name.

Shroyers:
George, 8;
Lucy, 9;
Gail, 6½

The Shroyer Family took their last trip into Corvallis in the fall. It was a train ride twenty miles to the East. Mabel's father followed his children to California; then he moved to Corvallis a few years after Oliver and Mabel arrived. Mabel always considered visits her father, Edward Hatch, and her step-mother, Clara Hatch, a delightful experience. New shoes for each child signaled the coming of the school year.

Summit School Class of 1915

CHAPTER 4

SCHOOL DAYS

Mabel insisted the children be photographed at Ball Studio while they were in town, visiting their grandparents, and preparing for the school year. The children were growing rapidly. Mabel wanted to send a picture to Tracy and Effie, her brother's family in California for Christmas. Eva, now sixteen, had finished the nine years of school available there. She was working as the postmistress in Summit, and was not able to make the trip into town with the family. Hazel and Ester had been writing, asking for a picture of the young Shroyer children as they had not seen them in four years.

George was nicknamed "Sober Sides" by his classmates. He earned this name in two ways. First, by his characteristic of seldom smiling. Second, he out-did everyone in a game the children often played at parties. They would do or say whatever they had to, to make you laugh! George never laughed! Everybody else would have to laugh, but not George, so they called him "Sober Sides."

> **Dad was principal of the Summit School the first year Lucy went to school. We got a new pair of shoes in the fall, before school started, and you had to take care of them, so they lasted you all winter. If our shoes needed any work, Dad would repair them. He put new soles on them, things like that. You couldn't get them too hot, or too close to the fire; it would burn them. You guarded your shoes since you had only one pair of new shoes a year.**

In the group photograph of the school's students, notice Oliver who is standing, in the center of the back row and somewhat taller than the students. Eva is also standing in the back row, furthest right. She has a big (dark) bow in her hair. Eva loved big bows. This was her style.
Mrs. Glyde Cone, their teacher, is standing to the far right of the students. Gail is sitting on the ground, second from the far right in the front row. Lucy is in the first row of children standing, second left of center. She is the darling little girl with braids, a perfectly straight part in her hair and a button down top with a perfectly folded white handkerchief in her pocket. George is also in the first row of children standing, second from the furthest left.
George always said his sister Lucy was pretty perfect. She was always right, and he could never get her into any trouble. Also, she was sure to tell their Dad whenever he and Gail were ever anything other than perfect.

The fall of 1912, found Oliver working as principal of a growing school. This was work for which had prepared himself. This was the work he had dreamed of doing and he was doing it.

In those days, about eighty children came into Summit to attend school each winter coming from the small farms in the surrounding area. That was an

exciting year, Lucy, was the first of the younger Shroyer children to start school. She and Eva, her older sister, were out the door early each morning with Oliver, their father. Occasionally, they rode horses to school during bad weather that first year, since Lucy's pace was slow. They were to spend many other early mornings walking from their home to the growing community of Summit, Oregon.

> Everybody had a bunch of kids; most of them walked or rode horseback to school. Lots of them came in from a-long-ways around. We walked two-and-a-half miles to school; most of them didn't have much further to go than we did. The Pettets lived quite aways-out, and they couldn't always get through in the winter time.

Oliver's role as principal of a school would involve him in the education of the children of this rural Oregon community. The school was sure to thrive under the guidance of this man. He soon began to enhance the Summit School. He built a large enclosed porch, at the entrance of the building, to provide additional protection from rain, and to give the children a place to play when the playground was too muddy. The community soon grew to appreciate the changes Mr. Shroyer initiated. When George and Gail started to school, they began helping their father in the mornings and evenings around the school.

> Gail and I had a job, before and after school, each day. The school board let us have the job because they liked Dad. He was a good principal. Somebody had to start the fires in the morning, pack in the wood and water. We had to sweep the school out every night and if we didn't get it finished, we had to finish it before school the next morning. Then each night, we had to clean off all the blackboards. At that time, there were blackboards all the way around three sides of the room, these big blackboards were about four-feet wide. We also had to dust all the erasers.
>
> Those were muddy days, the roads weren't paved, there was nothin' but mud, we didn't even have any gravel. We'd oil the floors once a month to keep the dust down. The oil came out of a five-gallon can; it would take about ten gallons of it, to oil the floors. Once a month we were paid, we got ten dollars between the two of us. After a couple of years they raised our pay to fifteen dollars, that was in 1917. That was the job we were paid for, we did that every year since I was about eight.
>
> We did other chores early each morning before we went to school. Dad started having me help him milk the cows when I was six years old. Lucy milked some, in the evenings, during the summertime. Gail never started milking until a year or two later. He never got along well with animals, the cows would chase him and dogs would bite him.

Oliver J. Shroyer, Principal.
He also built a manual-training building and taught the class.
George related there would not otherwise have been a shop if the school had a different principal.

When I got to be eight or nine years old, Gail and I would do the milking, clean out the barn, all those things by ourselves.

We didn't have any gym to play in but Dad built a manual training shop for the school in about 1918. It wasn't a very big building, probably thirty by fifty feet. He was responsible for teaching shop, so the kids and Dad built the shop, he just showed them what to do; they didn't hire any carpenters, you know.

We wouldn't have had the shop; if we'd had some other principal. I don't know whether we received a grade for shop or not, it was not a regular state requirement, I know that. At that time, you had to pass state examinations in English, history, and math, etc. You had an exam for each grade to take before you could get out of that grade; I did well enough to pass everything. In eighth grade you took your state exam. I barely passed English; did pretty good in history and geography, did real good in math and spelling.

We learned a lot in Dad's training shop; he taught the boys to be carpenters. There were small machines in there; we had a drill press, a lathe and things like that. Back then, before we had electricity, the machines were run by hand or by foot pedals. The kids liked that real well; they could learn how to build things. Some of the kids that got their lessons finished early; could go out and work in the shop; provided there was some work he'd already taught them to do.

It was only two or three years before Dad talked the school board into eight months of school instead of six. They had one man on the school board that didn't want the kids to have a shop at school and he didn't want more months of school; in fact, he didn't think kids ought to go to school at all. That was the only board member he had any trouble with. One of the other two board members was Underhill; he was real smart and helped a lot.

The Shroyer children each differed in their approach to school. The girls did well and seemed to enjoy it whereas, Gail didn't like school much, and George did only what he had to.

I didn't study much, but I always got passing grades; they couldn't understand that I could just listen to the teacher. At that time, they called school, class, the whole class went up in front of the schoolroom. They had one-great-big-long bench that you sat on, no place to write, just a place to sit down. Then they would ask you each, to go up to the blackboard, or maybe the whole class would go up to work on some math to see who could figure it out first. Then you had to read the same way, and my big trouble was that I stuttered, if I could get started reading, I could read and I didn't have any trouble reading. Sometimes I'd stutter and stammer around; I just couldn't get started reading at all.

We had to make up our own games in the old days; they encouraged us to play lots of running games. It was good for your lungs, your legs and everything. Some people were good at it and some weren't. The kids didn't have places to play like they do now. Although we did play croquet outside, that was quite a game. I don't remember ever playing baseball, instead we played Ante Over the School House, we'd throw a ball over the school house and someone on the other side would catch it. We played Blind Man's Bluff, Shinny, King of the Mountain, and marble games.

To play Shinny, we'd have a beat down tin can and we'd kick that around, that was kinda fun to play at recess. When we played King of the Mountain, we did that on a knoll out there in the school yard. The mountain was just this pile of dirt with grass on it. It was pret'near as big as a room and probably six-feet higher than the school yard. Whoever could get up there and keep everybody else off was king. You could do that by just pushing them off. Then you figured the stronger you were the more apt you would be to be King of the Mountain. I was pretty strong for my age, for the grade I was in; so I'd be King of the Mountain for my grade. That is one of the things that we played a lot.

We played a game with marbles. You'd draw a little diagram and everybody would put their marbles in there. Then we'd take turns shooting at other peoples marbles to try to knock them out. Whoever's marble you could knock out of the ring, you could keep. Gail and I were pretty good; between the two of us we had most all the marbles.

The reason Gail and I had so many was that you could buy ten little dopies for a penny. Those were just cheap little marbles made out of clay. They had a marble called a glassy, made of glass; they would be more expensive and they would be different colors. Then they had a "steely." It was made out of steel and those cost ten to twenty-five cents apiece. Those were awful expensive, Gail and I had a little money that we got from our janitor work so we could buy steelies. We used them as our shooter. Then sometimes, when you'd hit other peoples clay marbles they'd break and they wouldn't have that marble any more.

When it came time for recess to be over, they'd ring the bell for us to come in. We'd holler "grab stakes," that's when you could take anything. That's when Gail and I would usually get most of our marbles. I don't believe you ever put your shooter in, that would be your steely as a rule, if you could afford one. We just carried our marbles around in our pockets; some guys had little bags they put their marbles in. Most of the kids didn't have extra marbles. Gail and I thought we were pretty good; we had a big stock of them at home.

Then a couple of guys came in from South Dakota, they were in the same grade as Gail and I, but they weren't as big. I don't remember if they were as old or not. Those guys took all our marbles, they could play so much better then we could. Gail and I were running out of marbles before those guy's left. They didn't stay too long; we were sure glad to see them leave. When they left, they took all of our marbles with them.

There was one more game that we played; we called it Hounds and Deer. The property of the school didn't have over a couple of acres in it. There was a lot of private land and it was more open than it is today. There were mostly big trees; the brush had died out underneath them so you could see about a quarter of a mile away. Today you can't see very far because of the underbrush. We could run and play at noon; we'd have an hour anyway, so part of that time we'd play Hounds and Deer. The kids would go outside and there would be so many hounds and so many deer. The hounds would try to catch the deer. Some guys would be running and the others would try to catch them; when all the deer were caught, then the hounds won the game. If we were playing Hounds and Deer, the teachers would

Eva Shroyer (LEFT) became Postmaster at Summit at an early age and was courted by Jesse Kipper (RIGHT). They married on September 21, 1919.

ring the bell five minutes ahead of time. So, if we were a quarter of a mile off, we'd have a chance to get back in to school on time.

Dad saw that we had teeter-totters and swings, four or five of them. We had some for the little kids and then we had some for the older ones. Ernest Chapman, this half-Indian guy that went to school; we called him "Chap." He and Gail were always fighting, so when they got on the teeter-totter they would each promise that they wouldn't jump off. They would wait until they could get the other one up in the air, then they'd jump off. The other one would come down with a bang! It's a wonder they didn't kill each other.

I used to tease some so I got in trouble sometimes. I still remember one teacher; she'd use the ruler to hit you on your hands. Then, on the older boys she got to using the sharp edge of it on your hands, that really hurt.

This one girl, McCormick was her last name. She set right there in front of me at school; she was a good student. I never studied much, I

was always playing around. You got little prizes when you'd buy candy or cracker jacks and I'd gotten a little bull dog about an inch long. He had a pin on one end of him and you could stick that right into the toe of your shoe. So I did that. The seats of our chairs had about a four-inch crack in them. So, I stuck the toe of my shoe into that crack, and kicked her right in the butt! Boy did she yell the first time I did that! It'd scared her and she'd yelled right out!

Well, I started studying like a good guy. This teacher that used the ruler on us, came back there and asked; "What happened?" When this McCormick girl told her, I guarantee you the teacher whacked away on me for quite awhile. That afternoon when we got home from school, Lucy told Dad about that; so he gave me a whippin'. Dad was pretty strict but he never whipped me at school. I was real good whenever he was around.

When I was older, I got into trouble another time. We had one real strict teacher who taught the physical education classes that were started after the war. She'd start our class by saying; "Everyone arise, open the windows, (all of us next to a window were suppose to do that), take a deep breath, and throw your chest out." So one day, I did, I just jumped right out that window. Everyone but the teacher thought that was real funny. The teacher didn't think that was funny at all. I'm sure Lucy told Dad, cause he gave me a whippin' that night.

I remember one time the three of us kids had a real big surprise. When we got to school one morning the other kids said; "There is a new baby at your house." We couldn't believe it, but everybody said it was true. After school, we couldn't wait to get home. We ran all the way to see if the other kids were right. When we got there we had a new baby in the house. That was my little sister Marie.

In those days it was the custom that the parents didn't talk about a new baby coming. The year that happened, we were pretty young, I was nine. Not many kids today would ever have an experience like that. The three of us couldn't believe the other kids at school when they told us we had a new baby at home. Dad must have told one of the teachers that morning.

We kids were real excited about having a new baby in the house. Eva was seventeen years old and Lucy was ten years old so they got to help Mama with our new baby. I remember she was real little and delicate. I remember Mama and Eva made Marie some pretty little dresses. Eva always helped Mama make our clothes; most of those were made out of flour sacks.

> After that, it seems like Lucy took care of Marie most of the time in the summers. She thought that was lots of fun. I liked to play with Marie when she was a baby, I thought she was pretty cute. I thought it was fun having a baby in the house. I know Dad didn't like the idea; I heard him talking about it afterwards. I know he didn't think you ought to have more kids than you could afford.

Oliver was certainly ahead of his time. He realized that people needed to be responsible for the well being of their children. They should balance the number of children they raised with the amount of money they could earn. He believed that with the right to bear children came the financial responsibility of providing for their needs. Few people today, even now, understand that hunger and disease is nature's cruel weapon against man's inability to responsibly procreate.

Perhaps, Oliver had dreamed, before the birth of their last child, of being able to help his children through college. He turned fifty the year Marie was born. He was twelve years older than his wife Mabel; she was only thirty-eight. Marie was seventeen years younger than Eva, he'd already spent almost twenty years raising children before Marie was born.

> Christmas time was always something to remember; we had a program at school. We had that in the big room we called the "Study Room." Everybody brought lots of chairs in from the other rooms for that. Well, when we had the Christmas Program everybody was in it. After the program, Santa Claus would come in and every kid would get a sack of hard candy with an orange or an apple in it, hard candy was popular back then. I remember after the second or third year, we also had peanuts, we didn't have soft candy in those days. I guess the school board took up a collection so they could give us that for Christmas.

In early 1918 the Great Flu Epidemic, also known as the Spanish Flu, raged throughout the world. This devastating flu took the lives of millions of people. Eva was the first of the Shroyer family to come down with it. She was working at the post office as postmistress then. Jess Kipper the railroad depot agent who had been courting Eva, found her passed out on the railroad tracks one morning. He picked her up and carried her into the depot and waited for Mr. Hymes to make his return trip to the farm. He knew Mabel would take good care of Eva. Mabel was nursing her through this, when Oliver came down with it.

Oliver, not teaching since the school had closed, was working in the community helping care for those who had come down with this life-threatening flu. He, no doubt, had contracted it from those he had been caring for. Oliver became significantly worse each day. Mabel, becoming frightened, decided to send for the doctor from Philomath. She asked Mr. Hymes to telegraph Dr.

Newsom in Philomath from the Summit depot. The telegram read:

OLIVER SHROYER STOP SUMMIT OREGON STOP CRITICALLY ILL STOP MEET AT DEPOT STOP TOMORROW WITH HORSE STOP

In those days the train traveled from Corvallis to Toledo and back each day. When seeing patients in the country, the doctor took the train in the morning and returned late in the afternoon. This would be the very first time a doctor had ever come to the Shroyer's home. Oliver and Mabel, both trained in the medical profession, had dealt with previous medical conditions themselves. Mabel had given birth to Marie a year and a-half before with only Oliver's assistance. George was the only one in the family that was able to go to Summit. It would be his responsibility to take Trixie, their horse, to Summit to meet Dr. Newsom and guide him to the Shroyer's home. George, only ten, saddled up Trixie and rode two-and-a-half miles into Summit to meet the train from Philomath. When the train arrived, Dr. Newsom was surprised to see a bashful young boy waiting for him near the platform. On this cold blustery day Dr. Newsom rode Trixie out to the Shroyer's farm while George walked along ahead to show him the way.

> **I remember when Dad got sick, he stayed in this little room off our living room, it was warmer there than it was upstairs. Mama told us kids to stay away from him; she didn't want anyone else to get sick. Before the doctor came, I remember seeing Mama putting an onion poultice on Dad's chest. She covered it with hot steaming towels to keep it warm. That poultice turned bright green; we'd never seen anything like that before. The doctor brought a tonic and some pills with him, he gave those to Dad; they were giving people that for that flu. He wanted to know what Mama did for Dad and she told him about the onion poultice. He thought that was real good, so they put another poultice on Dad.**
>
> **Mama fed the doctor a good dinner before we started back to Summit. That sure seemed like a long two and a half mile walk that afternoon, it was pretty cold and real muddy. It was real late by the time I got back to the farm, it was pretty dark. That was the first time I'd been out alone after dark, that was kinda scary. I think the doctor coming must have helped Dad, he got better in a week. I felt good knowing I'd been able to help Dad. I remember Mama thought Dad had come close to dying. I just knew Mama and all us kids couldn't have made it without him. None of the rest of us caught it; they were saying it usually attacked the strongest and healthiest, young adults. I remember one family lost both their folks; their kids were on their own. That was really tough, I tell you!**

We had a pretty good school in Summit during the First World War. Dad was always principal and taught the higher grades. He was chosen by the Oregon Superintendent of Schools to act as one of three representatives to attend a special session on Education in Washington D. C. We only had eight grades at first, then after the "First World War" we had nine grades. We had more students in the whole school than Philomath had in their high school, twice as many or more. I'd say we had about one-hundred-thirty students.

CHAPTER 5

SUMMER TIME

Summer was a busy time for the Shroyer family. Mabel was a strong-minded pioneer woman who supported her husband in this move to Oregon. She was a mild mannered woman who always, even under the most trying circumstances, maintained a pleasant expression. She was kind and gentle when admonishing her children to mind their manners or mend their ways. She could often be heard singing while kneading bread in the kitchen.

She had a hereditary deafness that caused her to withdraw from most social contact. She did maintain social visits with Mrs. Underhill and Mrs. Hymes from the community, preferring to visit with them in their homes. She had inherited the early progressive deafness her father, Edward P. Hatch, had himself experienced.

Mabel preferred to spend her Sunday mornings preparing the noon meal for her family. Her deafness interfered with her enjoyment of the services held at the church in Summit on Sunday mornings. Once she remarked to one of her children, "You have no idea how funny it is to watch the minister's gestures but not hear him when he is speaking." It was quite a hike to church and there was the ever present mud. She chose to remain at home to prepare a special Sunday dinner for the family.

> **We had good breakfasts when I was growing up. We'd have mush, or hot cakes or French toast and then you'd have bacon or beef. There was always plenty of syrup and lots of homemade butter. I remember Mama and Eva making the butter, Mama was real good at making butter. They used this glass churn with a big wooden paddle in it, then they put the butter in a crock and kept it in the tiled spring. They always had buttermilk left over after they churned. They liked to cook with that buttermilk. Dad and some of the others liked to drink it, but I didn't. Mama didn't either, she drank coffee and tea and us kids drank lots of milk. Mama also made cottage cheese from clabbered milk. I sure liked that cottage cheese but I never liked that clabbered milk. I didn't like things that tasted spoiled, but it sure made good hot cakes.**
>
> **We didn't have as much bacon as most people. Dad was raised on bacon back in Indiana. He didn't like it too well. We only raised a pig every other year. As Dad thought that made plenty of bacon and ham for the family, he thought we could eat beef instead. One of my favorite foods was the corned-beef Dad and Mama put up. That is how they preserved some of our beef. They use to put it down in crocks, in some brine and let it soak for quite awhile. Then they added a pickling solution; it kept all winter that way. Every month or so,**

> Dad would take it all out and wipe it off, remove the scum from the pickling solution and boil it, then let it cool before replacing the meat in the crock.
>
> Dad also made hominy, another one of my favorite foods. He did that by taking the ashes out of the stove, putting them in a tub and adding water. When it would be strong enough to feel slick between your fingers then we added the corn and let it soak over night, the next morning we just rubbed the corn together with our hands and the husks would come right off. The corn being heavier, would sink to the bottom and the shell would float on top just like chaff. We'd skim this off. Then Mama stored that in a crock. You'd cook that Hominy in a kettle, or boil it if you wanted to or you fried it. Mama fried it, we liked it fried in butter and I could just live on that!

Oliver had begun teaching his sons to take responsibility at a very young age. The amount of work Oliver required his sons to do then would surprise us today. George and Gail began going out with Oliver to cut wood when they first moved to Summit, Oregon. During those early years he would cut up the wood, then George and Gail would place it in piles to dry. Little did Oliver realize he was giving George and Gail the training that would provide them with a stable income for the rest of their lives.

> On Saturdays and Sundays, during those first few years, Dad would fall and buck; and saw up all our wood. Dad always cut up the wood with a cross-cut saw; he'd also split it so it dried well. He'd even split it while it was out in the woods so it could dry real good during the summer time. Dad said, "Mama should never have to worry about wood." So soon as Gail and I were old enough, our job was to make sure the kindling and wood boxes were full.
>
> Dad first started teaching us how to saw up the small trees using the cross-cut saw when I was ten and Gail was eight. We often went out alone to cut the wood and stack it in piles so it would dry. Later when I was about twelve and Gail was ten, Dad showed us how to fall the trees. After that, Gail and I fell and bucked and sawed-up most of our wood. Dad would fall only the real big trees.
>
> In the summer we'd fall Alder trees, while they had their leaves on. Sometimes our cows would get right in our way when we were falling those Alders. I remember those old cows; they just loved those leaves, and we'd have to drive them away before we could fall the trees, you know.
>
> The wood dried all summer long, until after the first light rain, then we'd get all our wood in. We had that Alder all split and ready to put into the kitchen stove right out of the woods. Most of our heater wood was wild cherry; our heater stove took big chunks. We cut lots of wild

cherry in the early spring. We had most of it all cut up before school was out. In the fall we hauled in lots of wild cherry; I'll tell you.

I remember the time Dad went to Washington D. C., to represent the State of Oregon at a special congressional session on Education. He was going to be away for about six weeks. He was going to visit relatives in Indiana after the conference. Dad told Gail and I to cut the winter wood while he was away. We were only thirteen and fourteen years old then.

Gail and I knew he didn't want us to buy a drag saw because they were real dangerous. They weighed probably three-hundred-and-fifty pounds and they ran on a gas engine. The drag saw rested on the log, but had its feet on the ground to support the weight of the saw. We were looking for an easier way to cut the wood and we found a drag saw for seventeen-dollars-and-fifty cents so we cut enough wood to pay for the saw. We cut cords, and cords of wood that summer; enough for us and for the neighbors. The "Old Man" came in a day earlier than he'd said he would, and we hadn't gotten rid of our saw yet. Boy! Did we know we were in trouble, instead Dad said; "Perhaps I was wrong. You did a good job, paid for your saw, were careful and didn't get hurt. How could I be upset with you?" We thought that was real fair of him.

The next summer, when Gail and I were fourteen and fifteen we decided the "Old Man" was too strict so we decided to leave home, that we could do better without him. We found a vacant house about two miles away. We took some of our things and we moved over there. One evening we saw Dad lookin' in the window. He had walked four miles round trip through the dark to check on us. After a few days we got hungry and we went back home.

Dad always wanted us to have all our wood cut up early in the year because we needed to start planting our garden pretty early. Our family always had a real good garden. When Gail and I were big enough we'd hoe the garden and pick the berries; we had great strawberries. In the fall we had squash, cabbage, broccoli, potatoes, and onions. Dad helped us since Mama was a city woman and didn't like gardening much. We had chickens, a pig or two, four or five beef, and a cow or two. We had a team of horses and a saddle horse or two.

In late February or early March, they would plant their peas. This meant they could have peas early in June. The family garden including the orchard, berry patch and vegetable garden grew to better than an acre. Their garden area was always at least two-hundred by three-hundred feet. It took a great deal of work to care for it. No irrigation was required since the weather in Summit never became hot and dry in the summertime like it did in Corvallis. One day

Mabel called Oliver to the window and there they watched George and Gail hoeing the garden. Gail was working hard and fast, putting lots of energy into his work. George was barely moving, half-heartedly working. It seemed that George was thinking to himself; "No use to hoe too fast, just leave it for the ones that really like it!"

> Marie recalls her father saying of the boys. "Gail is a hard worker; he will do all right in life. George will get somebody else to do his work for him; so he will be all right too." Once George started working for money all that changed. He worked real hard.

Early June found the children out in the berry patch almost daily. They always had a strawberry patch and there also were raspberries and logan berries, later came the goose berries and currants. Over in the orchard, the children picked the wild plums first. I'm sure there is nothing much better with homemade bread than wild-plum butter. Next came the crabapples. Mabel and Eva always made pickled crabapples to serve with the meat. The crabapple tree stood behind the house.

One hot summer day, the preacher's son unexpectedly came to the house to visit Lucy. She saw him coming in the distance and ducked around the corner of the house just as he pulled up. Embarrassed that he might see her without her shoes on, she crawled up the crabapple tree behind the house to get on the roof. Then, she ran over and climbed into her open bedroom window. All this so she could come casually down the stairs, with her shoes on.

The orchard provided the family with plums, pears, and apples. One variety of apple made great applesauce. Another apple was good for canning, making pies and drying. They also had winter apples that they stored in a cool pantry. These lasted throughout the winter months, providing fresh crisp apples for the family. The children liked finding those in their lunch boxes.

Peaches wouldn't grow up there in the Summit area, so Oliver never failed to buy a couple bushel of them. Mabel and Eva put up peach preserves and canned the rest for winter. Eva was good help; Mabel got along with Eva even better than she did with Lucy. Lucy tended to be a little bossy but no one seems to know where she got that trait.

> Mama and the girls had all that canning to do; they'd open the doors and windows to cool the kitchen. Sure smelled good sometimes, when we'd come in for dinner. I remember our floors were bare, we didn't have any rugs or linoleum like our Grandma Hatch did. That made cleaning up after canning easier. Mama would scrub the floors with a big mop. She had a bucket with a wringer on it; that seemed to work real good.

The vegetable garden had most of the common vegetables growing in it. With one exception: This garden had no cucumbers. Oliver thought the cucumber

Lucy was a modest teen and when the preacher's son came to call, she avoided him until she could put on some shoes.

George's grandmother Hatch had an outdoor kitchen. Use of an outside stove avoided overheating the house during summer.

About 1918. (BACK ROW) Oliver's three sisters: Mary, Clara, Amanda; Mabel and Oliver. (FRONT ROW) Marie, Lucy, Gail, Eva, George.

was not worth growing, so they didn't make pickles. Oliver was also hesitant to eat tomatoes. When he was growing up in Indiana, people believed tomatoes or "Love Apples," as they called them, to be poisonous. Any shiny red fruit with a stem end blossom could be poisonous. Oliver did plant a few tomatoes each year, but they rarely ripened. They would pull up the vines and hang them in a shed, in hopes that a few tomatoes would ripen. The Shroyer family did not eat many tomatoes. Oliver planted corn every year also, rarely did it ripen since the summers were not long enough up there in Summit – elevation 729 feet.

In those days, the Fourth of July celebration was a big event. The community looked forward to this day when they could meet with friends and enjoy the entertainment of the day while relaxing and visiting.

> The whole town and all the people from the farms around would go into Summit for the Fourth of July picnic. There would be all sorts of different events The events and contests were first. They had wrestling matches, shooting matches and lots of races; there was always the three legged race.

The "three-legged race" was lots of fun; two people would stand side by side and they would tie your inside legs together. One tie would be above your knee, the other below your knee and you'd put one arm around the other guy, then you'd run together. Sometimes you'd fall down -- but you better not! The idea, is to take big steps -- right together, Gail and I practiced a lot at home and we were about the same size so we got so we could run pretty fast! We always, won that three legged race!

Once in a while we'd win a running race, those were run among certain ages. Different groups of certain ages would run, the boys raced first then the girls raced, after that the fastest boy and the fastest girl raced.

Last of all, they had a pie eating contest and you would usually get a blackberry pie. I never liked that messy stuff. There was a guy by the name of Hamer up there; he always won that. He was a nice enough guy, but everybody made fun of him with pie all over his face. Most older people wouldn't do that, they didn't want to get messed up.

After all the contests were over, we'd all go up in front of the Community Hall there in Summit. To find it now, you go clear through Summit; it's the last building on the left hand side just before you start down the hill. They use to have a little park in there. They didn't have picnic tables in those days so we all helped by bringing in lots of wood planks. The men brought their saw horses, to lay the planks across; that's how they made the tables. The chairs were inside the building.

We had lots to eat at those picnics. From then on we'd eat until everyone went home. What I liked most was the homemade ice cream. The men would make the ice cream right there. They would pack the ice in sawdust and bring it in on the train then they used that ice to freeze the ice cream. We didn't have much in those days; not the way we do today. We got by pretty good and had lots of fun; those were happy times.

Once the Fourth of July Celebration was over, it signaled the beginning of harvest time and the push to get the hay in before the rains came. This was a critical time.

Haying would start after the first of July. We'd cut it and after it dried, we'd start getting it out of the fields. Gail and I would mow the dried hay into the barn. After the first couple of times, Dad realized that I was having hay fever. After that, Gail did the mowing and Dad had me drive the horses while he set the hay fork. Each summer, we would get the hay in the barn by the end of July.

Oliver Shroyer with his rake. Not only was he a superior school teacher and administrator, he was a superior farmer. His 1918 Model T Ford at left.

When Dad was using the horses out there in the fields, he'd take the harness off the team of horses at noon and tie them up to let them eat. Gail liked to tease the animals so then he would go out and jump on this one horse and get him a stick and make the horse run around in a circle. Whenever Gail did that, he usually got into trouble. I remember Dad making him get off, saying that the horse needed to eat and rest.

After we cut the hay, Dad liked to turn the cattle in to eat the hay that grew around the fence rows or didn't get picked up. We had an acre of potatoes in there in one corner of the field and we didn't have it fenced, so Gail and I had to keep the horses and cattle out of there. I remember Pet, one of our horses, had a colt with her that year. She and her colt liked those potatoes real well. Gail and I were supposed to take turns watching that field and we thought that wasn't much fun. We thought it would be lots more fun if we'd both watch the horses for awhile, then we'd both go play together for awhile. Gail and I were having a real good time playing out there. It was Gail's turn to watch the potato patch and someone yelled that the horses were going in the hay and potatoes. Gail had a big long stick; he ran over there and hit this mare Pet, with that. I remember he hit her on the butt with that stick. She backed up quick, kicked him on the right side of his face, and knocked him out.

That was the year Dad was doing the finishing work on the house. So, I ran down there and got Dad. I remember him running up there, getting Gail and running back down to the house. Dad laid Gail on the bed; it was quite awhile before he came to. We didn't have any doctor for him so Dad sewed and taped him up as good as he could. Gail had quite a scar on the side of his face from that. Dad was real tough on us for doin' that. From right-then-and-there, one of us stayed at the house until the other one came up out of the potato patch.

Oliver took extra special care to fertilize the fields with cow manure after haying. After the hay was in, we started getting the manure out of the barn before the winter rains set in. Dad had built a manure pit by surrounding a hole with large logs. This pit was behind the barn; there were holes in the barn behind the horses stalls. We pitched the manure out into the pit through these holes. We had to get it all cleaned out of the cows and horses pens. We'd go barefoot to keep from ruining our shoes. I didn't like having manure on my hands or my feet. The horse manure wasn't so bad, it had more straw in it. But that cow manure, it was more manure than straw and I didn't like walking around in that. Gail didn't seem to mind for some reason, maybe he just didn't take on so much.

In the fall we would fertilize the fields by putting that manure into a cart. We had an old-iron-wheeled cart that was low-down to the ground. Then Gail and I would drive the team and cart about while we shoveled it out. After we'd finished, Dad would let us boys drive the team about, to show-off. You could put the manure out any time in August, September, or even October if the weather was right. We sure worked hard out there in the fields. Later in the fall, after the final harvest in the garden, we fertilized it with chicken manure.

Dad, Gail and I cleared land whenever we could. This meant brush burning in the fall of the year. Dad and Gail liked to burn the piles of dry brush in the fall. I never liked burning because it bothered my hay fever. After Dad saw that it bothered me, made my hay fever act up, he never asked me to do that so much. On Saturdays and Sundays, we'd work to clear more land whenever we could. That was the way we cleared our forty acres. Lots of the logs were old and rotten so you'd take a grub hoe and dig them out and pile them up so they would dry. We'd pile the wood on a stump any time we could. We'd cut it up into four to six foot long sections; split it up and pile it around the stump. Then in the fall of the year we'd pile up lots of brush on top of it and we'd set it on fire. Boy! Would that make a hot fire when it dried out. It would burn right up.

During the First World War we used some of our trees to make ship knees, these were used in building the big transport "Liberty Ships."

Dad, Gail and I could dig one out in a day. We'd dig out the roots til the tree would fall down, then cut off the tree about six-to-eight feet above the root and trim it and we'd have a piece, the shape of "a-large L." Most were a foot-or-more in diameter, they paid you more for an eighteen inch width piece than they would a twelve inch width piece. We were paid one dollar per inch. A buyer would come down from the shipyard in Portland and buy these from all the farmers around. I remember seeing five train-car loads leaving Summit one day.

Mama and the girls worked inside and they worked real hard all the time too. Doing the washing was one of the most difficult household chores. Mama did the washing by hand for the first two years then we got a washing machine, a wooden one. It had a lever that you pushed back and forth and it had a couple of scrub boards that scrubbed back and forth, but no wringer. You had to wring it all out by hand. To do the washing, you'd have to heat your water on the stove all day. Boy! Would it get hot in the kitchen in the summer time. It was awfully hard for women. I remember Mama was never cross not even on wash day. She had a good sense of humor, she used it to have a little fun when we were all hot and tired.

Whenever Mama was heating up the water to do the washing she'd put a big pot of white beans on the back of the stove and the beans would cook all day long. Then she'd make cornbread to go with the beans, cause Dad liked southern food. Next day, she would fix baked beans with the left over white beans, we sure liked that. Mama talked about Grandma Hatch's out-door kitchen, when her kitchen would be too hot. She thought that was a good idea and she sure wished she had one. Of course it was cooler in Summit than it was in Corvallis.

In the summer time Mama and the girls made all our soap. To make it they added lye to beef fat. Mama's stove was a good strong one – it was real sturdy. It had a water reservoir built into it. Lots of people didn't have a good stove, they just had a tin stove. You didn't dare put much water on a tin stove, cause if it got too hot, it would just buckle down, and you would lose your stove.

Mabel was a capable woman, well trained in the art of sewing, mending and quilting. She believed that a "stitch in-time saved nine." She kept up on her mending and trained her daughters likewise. She and Eva made almost all the clothes worn by the family. Oliver always bought the boys woolen underwear for winter each year and ordered most of the pants from Sears. The girls helped their mother make the dish towels and other household items. There was only one curtain in the house, other than in the parlor. That was a little white one just above the kitchen window. Lace curtains were hanging, only in the rarely used parlor; the remaining windows had only pull down blinds made of a dark green oil cloth material.

Many of the clothes were made from muslin flour sacks in those days. Mama made Lucy some bloomers out of those. When the clothes were hanging on the clothes line to dry, right across the butt in plain letters Lucy's bloomers said; "Royal Crown." This was the brand of flour we always bought.

Marie said; 'I remember Mama making me a dress from those muslin flour sacks. The dress was a light cream color, very soft, it had a neat little pinafore front. Mama had carefully embroidered a darling little teddy bear on it, this covered up the words "Royal Crown." The belt was attached to this and tied in back. This was my favorite dress; I wore it as long as I could.'

Marie also remembers that any time her Grandpa Hatch would come to visit, he always took a look at her mother's sewing machine. He made sure it was in perfect working order when he left. It was a treadle sewing machine from Sears. It was a "Minnesota" brand sewing machine and had been designed with a unique cover. Three sides were always protected by the sides of the box that held it. When you opened the top, that would cause the head to come up into the box, ready for use. You only had to let down the front side of this box. When you closed it, the sewing head hung upside down below the box. This made it very safe with small children in the house.

The summer before the First World War ended, Dad's three sisters came to visit. Mary and Amanda, Dad's half-sisters came from Indiana and Clara Shroyer, Dad's full sister came from California. Clara didn't ever marry. She was a missionary and worked with the poor in Los Angeles.

George remembers participating, only once, in a ritual that took place in the late summer. The three younger Shroyer children took the train into Corvallis to stay with grandparents. They were going to try to earn some extra money that summer.

There were hops fields near Salem and all through that country; it was one of the main crops in the Willamette Valley. Hundreds of Siletz Indians would come through this area in the fall of the year, on their way to pick hops. They would come over from Siletz on the main road from the coast. They would walk along this road every fall, the families living along there, would see them.

We picked hops at a field near Corvallis, the hops grew stretched out on wires; up on big poles. The men would have to let them down so people could reach them, it took several men to do that. They moved along on a wagon with a big platform built on it and they would be up in the air, they used long poles to lift the wires off the top and place them about six feet off the ground so you could pick the hops off the

vines. Some of those rows would be a mile long. Many of the fields were over one-thousand acres. Hops poles were about twelve feet high and had about an eight-inch diameter. The baskets you picked into were three-feet tall with a thirty-inch diameter at the top that tapered toward the bottom. They were made out of light wood.

The hops were about one and a-half inches long and three-quarters of an-inch in diameter. To fill a burlap sack, it seems as if it took two baskets full of hops. You weren't supposed to have any leaves in your basket. Oh, you could get by with a leaf in them occasionally but some people would just rake everything in there. When your sack was full you would take it up and have it checked. They would make you clean it if it had too many leaves in it. You tied the burlap sack closed tight with twine and then they would give you a ticket. When you finished at night, you turned those little slips of paper in and they gave you money for them.

I worked picking hops only one year. I didn't like it. Lucy was getting to be a young lady and lot of the men would help her pick hops. I couldn't keep up with her because there wasn't anybody wanting to help me. With all the help she got from the young men, she made five dollars a day and I made fifty cents. That's about the way it went.

I was a growing boy and I liked to eat something. Grandma Hatch just served us boiled potatoes with skins on them and sliced tomatoes. I hated tomatoes, and I didn't like to have to peel my own potatoes. Mama never fixed boiled potatoes with skins on. Lucy loved sliced tomatoes and so did Grandma. so they got along pretty good. I don't think she charged us anything for board. She shouldn't with that kinda stuff, I never went in to pick hops again.

The next summer, I stayed up at the farm all summer. I didn't even go into Summit. Mama, Lucy and Gail all went in and picked hops late that summer. That was when Marie was two years old. I remember Marie set out on the wagon while Dad and I hauled wood in. Eva was working as postmistress and at night she would be home to cook supper. Mama and Lucy, when they went in, they made a little money. Gail and Mama stayed about three weeks that time. Lucy moved in with Grandma Hatch then to finish high school in Corvallis. Dad wanted her to go to the high school there because she wanted to be a teacher. He thought she'd get a better education in a bigger school. It's a good thing Lucy liked potatoes with skins on them and sliced tomatoes!

In the fall and the spring when Gail and I were older we would walk four miles to Nashville every day for a week or two when the salmon were running. We'd spear us a couple or more salmon apiece. Mama

> canned them up. Boy! That was good in the winter. Marie remembers watching for the boys to come home from fishing. They had a long stick that they carried between them. On the stick hung the salmon they had caught that day. Often these fish were ten pounds or better. Quite a lot of weight to carry four miles.
>
> In the fall Mama canned a good deal of beef too. The remainder was put down in brine. There weren't many deer around then so we never went hunting or anything like that. Dad never took the time to hunt for deer. With everybody else hunting, deer were scarce. We could go for a month and never see a deer track.

Both Mabel and Oliver were good role models for the children. They were also an important force in the community as a whole. They kept up with new developments in the world and new ideas. Reading was one of their forms of entertainment. Mabel subscribed to the *Ladies Home Journal* and *The Christian Science Monitor.* Oliver subscribed to the *National Geographic* and a newspaper called the *Pathfinder*. They apparently agreed it was money well spent as these magazines provided their children with a glimpse of the outside world through their parents' eyes. I'm sure there were lengthy discussions in the late evening when the house was quiet, after the children were in bed. Mabel was sure to be working on the mending and Oliver was keeping up on the news of the war. He had paid dearly, no doubt, for the Philco radio he'd listen to each night.

The walls displayed a photo of "Oliver's Class at Valparaiso University," his "Business School Diploma" and his "Teaching Certificate." There were also hand painted pictures of horses. One was of a powerful horse. Another was of a black horse and a white horse standing quietly together in a field with their necks crossed. Lace curtains covered the windows in the rarely used parlor. This room held an ornate pump organ, comfortable chairs and the only carpet in the house. Oliver being musical, sometimes played the organ while the children fell asleep at night. He could play many tunes all by ear.

The "Minnesota" sewing machine operated by a foot treadle on the floor of the cabinet.

CHAPTER 6

LEAVING HOME

When George turned fifteen in 1922, his life took an unexpected direction. Now Oliver, as we know, valued education. Oliver, well educated, expected his sons to follow in his footsteps and to attain a scholastic standing equal to his own. But he hadn't realized the impact the pioneering life style would have on them.

George completed the nine grades available in Summit that year. Oliver recognized that, were his son to continue his education, he would have to live elsewhere. George, a shy young man, had not given much thought to what he would do about furthering his education. Perhaps he assumed it would be his lot to continue to assist his father with the family farm.

> **Dad told me it was time for me to leave home the summer I was only fifteen years old. That was quite a shock. I was pretty bashful and I hated to go out to look for work. I remember my Dad sent me over to Devitt to see about getting a job in the woods the year before. They told me I was too young to work and I was sure glad to find that out. Dad wanted me to work during the summer and then go to school in the winter. He always wanted you to go to school. At that time, there weren't too many people getting through high school.**
>
> **Dad told me I would need to make my own way in the world once I moved into Philomath to start the tenth grade. I suppose since Dad had been on his own since he was thirteen years old, he figured I would do fine. I was never home much after that. For the next two years I put in most of the fire wood for the high school, and I mowed lawns that first summer. I did what ever work I could find. That first year, I stayed with Eva and went to school. I always called Eva's place the "rooming house" even though I was the only one staying there. Nobody else stayed there, just me. Once, Eva and Jess took a trip back to Missouri to see Jesse's family. That was the first trip they had taken since they were married. They asked me if I would take care of the house, look after things.**
>
> **Eva said I was welcome to eat the pear preserves; she knew I loved her pear preserves. We'd never had any at our house; I pret'near ate them up. She also had lots of canned salmon; I sure liked that. I'm sure she didn't really mind, and Jess never said anything, I don't think he would have ever said anything. I had the feeling that they thought I was eating lots, but, I was growing up pretty fast.**
>
> **One morning when I walked to school from Eva's house there was a bunch of students standing around. When I got there, I saw a dummy of principal Bump hanging in effigy. The other kids were standing there staring at that, but they weren't laughing, I thought that was**

real funny and I started laughing. The principal came up the sidewalk about then and thought I was responsible for that. Principal Wilson Bump took me in his office and expelled me. His son was my age so he must have got the clothes for it. I had to get Dad to come into town to help me get back into school. I told him I didn't have anything to do with that doin's. Dad knew I would never lie, so he talked to the principal and got me back into school.

George had been known as "Sober Sides" during his earlier years in school. Life would no doubt change for him, although it would not be evident for some time. While only fifteen, he was experiencing his first year away from home. He had not yet gained the full degree of humor he would be known for later in his life. Only subtle glimpses of this young man's quick thinking and skillful reasoning were beginning to surface.

I had some pretty good friends in high school, H. D. Moreland was one of those. H. D. was a good friend of mine – the banker's son. He was pretty smart in high school. I ran around with him quite a bit that first year. His mother was real good to me. As I had no folks in the area, I'd go over to H. D.'s. They had a big finished basement that was warm and dry, H. D. set up a radio repair shop in there. He'd repair people's radios, as most people didn't know anything about them.

Dad got one of the first radios in the county It had earphones on it so you could listen to it. Dad took it in to H. D. and had him separate the earphones so two people could listen at one time. I remember he always listened to it in the evenings in the parlor after we kids went to bed. It ran on batteries in those days. Dad always brought his radio in to H. D. for him to work on it whenever something went wrong. Radios weren't made too well in those days as the makers were just figuring how to make them. Most folks didn't have telephones or radios then.

H. D. Moreland and I use to go to the movies on Saturday nights with the Moreland's. They ask me to go with them lots of times and I always had a lot of fun with them. They were real nice. There was a time when short knickers were in style and they offered to buy me a suit of those once. They thought my Dad would pay for them when he came in. I knew Dad would not pay for my clothes, that he expected me to buy my own. That's why I kinda quit going over to Moreland's. It embarrassed me that they wanted to buy me something and I never had money to pay them back. H. D. Moreland became an electrical engineer and did real well for himself later in life. He got a job installing the first x-ray's that came into this country. He installed the ones over at the Corvallis General Hospital. He made good money during his life.

I only had one pair of pants to go to school in. In those days, you didn't wear overalls to school. You wore wool khaki pants, kinda like the ones that soldiers wore. I was responsible for washing my own clothes since I didn't live at home with Mama anymore. I'd wash and dry those on weekends, then I'd wear them to school during the week. One time I found out that a little lye in the water would help them look a lot cleaner. While Eva and Jess were on vacation, I got too much lye in there and didn't get it mixed up good. When those pants came out that time, most of one leg was off! I couldn't wear my khaki pants to school any more. School was about out; I remember that. I can't remember what I wore to school the remainder of that year. I may have just worn my waist overalls to school for the rest of the time. When I quit running around with H. D. so much, I got more acquainted with Bill Hunter, another guy in my class, and we started running around together.

Toward the end of my sophomore year, Bill and I became best friends; he was kind of a "country bumpkin" and so was I. That's when I met his sister, Helen. He and his sister came in to school from Wren, in their car. On weekends Bill and Helen would want to go to the movies and I'd go along. I remember that next summer, Gail and I ended up staying at the Hunter farm. Mr. Hunter never charged us for our food when we stayed with him as he was very generous. Gail was going to go to high school there in Philomath that next winter so he'd left home. That was the summer of 1924. Mr. Hunter wanted to build a barn there on his farm that summer so Gail said he would help him. Gail would work on the barn while Bill and I went to work with Mr. Hunter who was building a mill for Mr. McCoy over in Kings Valley that summer. We all worked on the barn in the evenings and on weekends. Helen, his daughter, was only sixteen then and she did all the cooking for us. She was an awfully good cook. She was pretty busy that summer feeding all of us. Her mother had died a couple of years before so she had lots to do around the farm.

Helen's parents, William George and Rosa Nell Hunter, moved here with their children Bill and Helen, from Mosier, Oregon in 1913. They'd lived out by Mount Baldy when Helen was a little girl. Her father had been a foreman for a work crew on the Northern Pacific Railroad. Later, they bought this forty acre farm near Wren where they had built a large chicken farm before George met them.

Helen said; 'I had been working hard all day helping my neighbor scrub and iron. I'd come home tired, but had fixed some meat for supper and baked a cake. That night Bill brought George home with him for the first time. After George went home I told Bill; 'You get

(TOP) **William George Hunter and Rosa Nell Prather Hunter, parents of Helen.** (LOWER) **Summer of 1924. Helen as she appeared the summer George Shroyer lived in the Hunter farm while he worked at McCoy's Mill. After a first meeting, Helen declared to her brother, "You get him out of here. I don't want to cook for him. Later, I married him, cooked for him and raised his children."**

him out of here; I don't want to cook for him'! Later, I married him, cooked for him and raised his children.

'My Mama did most of the work required, for our chicken farm. Now, I had most of the responsibility of caring for the chickens; we had about three-hundred. The chicken house was a long building, perhaps as much as one-hundred and twenty feet long. There were three levels of roosts for the chickens to set on at night. The nest boxes were all along one wall. The chickens had to have food and water; and their eggs gathered daily. Once a week I had to take the eggs to Wren, two miles away, where I sold them to the general store. I always packed, loaded and unloaded the eggs by myself. I drove the horse and buggy carefully loaded with all those eggs. From Wren I continued on to Philomath, five miles away, to buy feed for the chickens and animals. While there, I bought the family's groceries. I was very fearful of the trip each week.

'The horse that pulled the wagon was Jess, a dapple-gray Clydesdale draft horse. My Papa broke him after he brought him to Benton county. Jess had a star in his forehead and a Roman nose. Two times along the way the route crossed the railroad track, I would always hope that we would not meet a train at the crossing. Jess was scared of trains and cars, they were very noisy in those days and he was easily frightened by them. He was such a big horse he was hard to handle when he was scared.

'In high school, I took one year of cooking and one year of sewing. These skills helped me take over the responsibilities of the home that were left to me. I canned about two-hundred jars of fruits and vegetables each year while I was a teenager. There was a large patch of wild strawberries on our farm. One year I canned twenty quarts of wild strawberries. I use to pick gooseberries from our tamed gooseberry bushes. Papa, Bill and George were fond of the gooseberry pies I made.

'I was only fifteen years old when I lost my Mama. I made sure not to let her plants die. She had beautiful house plants and flowers filled the yard. My fondest memories are of my Mama out working in her flower garden. She dearly loved her flowers; she always had a lovely smile when she brought flowers in for the house. Aunt Agnes, Papa's sister, liked flowers too. I remember Papa always said that he never got to live in a house without flowers all over the windows. I've always loved flowers since I was young. They are a special part of my life.

Papa and Bill were going to be living at Mac McCoy's mill site that winter so they could finish building the mill. My brother decided he'd rather help Papa build McCoy's mill than finish school. They moved

George Shroyer's 1923 Model T Ford Roadster

out to Kings Valley and lived in one of the mill shacks there and ate in the cookhouse with the other men that winter.

I wanted to finish high school so I needed to stay someplace in town during the winter. Originally there were two families, a Pettit family and a Pettit family in the Summit area. The Pettet family lived on a farm about five miles away from Summit, whose children went to school with George. Mr. and Mrs. Pettit ran a boarding house in Summit before they moved to Philomath. There, they ran another boarding house, and a cookhouse for the railroad crew. Papa paid Mrs. Pettit fifteen dollars a month for my room and board. I helped her so much; she didn't charge me anything for the last month. She said; "I was the nicest girl that ever lived in her house." I always felt this quite a compliment considering Mrs. Pettit had grand daughters also staying with her.

George said; 'Dad always wanted Gail and I to go to school in the winter so that's what we did. We figured he knew what was best for us.' That was the first year Gail came into town to go to high school. We rented the old Lutz place and we called it the boarding house. There's hardly anybody alive that remembers that old place. It was close to Eva's and the school house. The; the school house was over where the grocery store is now. Gail and I spent that winter living at the Lutz house so we could go to high school.

One Saturday night late that fall, Bill came in with his pickup, Gail and I were running around town with him when we noticed this party going on. We could tell that they were having quite a bit to drink from an old keg out on their back porch. We decided we'd swipe that keg of "apple-jack" so we took it home. Chuck McDowell, Cecil and Emmit Green came over later so they had some of that apple jack with us. Bill and Chuck had a lot of that. Gail had some but I didn't have much. After awhile we were all trying to go to sleep. Bill kept on singing this song and tapping his foot – tap tap tap tap. He could sing about like I do. Chuck, he jumps up and grabs an ax. He stood over Bill's head yelling; "There's goin' to be another fiddler in hell to-night." Gail, he'd been around drunks before, so he just grabbed that ax before Chuck could swing it. That could have been real bad, you know. I decided that my Old Man knew what he was talking about, that we shouldn't be playin' around with liquor.

I had lots of different jobs when I was in high school. At first, during my junior year at school, I worked for the railroad on Saturdays and Sundays. They had me take a three-wheel speeder-car and ride from Philomath on up above Harris. Somebody else would come up the other way from Nashville. The speeder is the car that checks for safety, you could tell if the rails were OK, if not, your speeder would run off the track. It never ran off the track when I was working at that. The railroad paid me by the day, I think I got about three dollars a day for doing that.

Pretty soon I got to working for Rex Clemens and Joe Rodgers. I loaded railroad cars for them on Saturdays and Sundays while I was in my Junior and Senior year at school. Rex Clemens might have been a railroad man if he hadn't had an accident on a speeder. One day he met a freight train between Wren and Harris. He jumped off, leaving the speeder on the track, and ran toward the train to keep the speeder from hitting him when the train hit it. This probably saved his life. After that he walked into Wren and got a job with Joe Rodgers loading railroad cars. Soon after that, Joe Rodgers took Rex into business with him. This was Rex's first start in the lumber business.

The following summer, I moved back to the Hunter farm and Gail stayed with Eva and Jess. Mr. Hunter was very happy to have me around as I was real helpful, you know. During that summer I got a job up at Kings Valley again because Mr. Hunter had just finished building the mill. McCoy's new sawmill was the biggest saw mill in the country. I ran a cut off saw for awhile but I didn't like that because I got sawdust down my neck. It wasn't long before I got the job of loading the railroad cars with lumber.

Bill and I, and Mr. Hunter, went to work in his pickup, I usually drove, Bill never felt like driving in the mornings. The roads were full of mud holes, they didn't have near enough rock on them, it was mostly river rock. There were lots of holes and you wouldn't know which way to go in those mud puddles. I got to noticing Mr. Hunter's fingers and thumb, he'd always turn them to the right or left, which ever way he would go to miss those ruts. So, I'd go that way, Mr. Hunter thought I was pretty smart. Bill, he never noticed that, and I never told him. He'd always drive the wrong way; so Mr. Hunter was always cussin' him. He'd say; "Why can't you drive like George? He can tell which way to go."

During my last winter I was in high school, I lived at the Hunter farm with them. Bill had quit school, so, I drove Helen and myself back and forth to school in Mr. Hunter car. He had an old Chevrolet then, I think they called it a 490. Later, Mr. Hunter traded it off and got him a Model-T Roadster.

One Sunday we went for a ride in his roadster. We were coming back and we were just above the Wren school house. We had this little dog, that sat on Mr. Hunter's lap, who would put his front feet on the door so he could see out. The Model-T Ford, it was real low. When going through a mud puddle once, I took a sharp turn to the left and that dog fell right out in the mud. Old Mr. Hunter, he yelled; "Hey! Hold it! Hold it! Hold it! The dog fell out!" He says; "The dog fell out"!

So, we stopped and got the dog. Mr. Hunter took him down to the creek and washed him all off. That dog, when he got back in, kept all four feet on Mr. Hunter's lap, and just barely stuck his nose out over the door. He'd stick his nose around and look, didn't want to look through the window. That's how he came to fall out in the first place. He'd put his feet up there and he'd look all around. He'd just kinda teeter up there on that darn door – that taught him to stay back!

George and his new friends certainly had some fun times during those years. He was beginning to relax and enjoy life some. He was a hard worker and felt it was his responsibility to apply himself at any work he did. School was somewhat less important to him.

I got my first car while I was still going to school during my senior year. I was real tickled with that car. I wasn't eighteen yet, because I had to go up and get my Dad to sign the papers on it. It cost one hundred and seventy dollars and I paid seventy down I remember and I owed one hundred.

I had to have Dad sign it so they would accept it. I went up Sunday, the following weekend to ask him if he would sign it. He said; "No, he wouldn't sign anything like that." I remember that scared me, I thought I'd lost my money and would lose my car. He told me he was going in on Monday and he would pay it off. Dad told me I could pay him fifty dollars a month for two months and he wouldn't charge me any interest. I got my car in a hurry and got it paid for right away. The interest was high and it would have doubled the money I owed.

That was an important lesson in financing, one that I remembered. Before, I didn't care whether I had the money or not, as long as I got the car. I learned that you had to save so you could pay cash in order to keep from paying interest. Helen told me she was glad that I was learning to save money. She learned in her economics class that if you ever wanted to get ahead in this world you had to save part of what you made. "That was the most important economics lesson she had."

That car was a Ford Roadster and they called it "Bugs." Every one of them was cut down; to make them into speedsters of some sort. It had lots of extras on it. I figured it would out-run any car in the country. After that, Helen use to holler at me; "Don't drive so fast!" We were studying about Greek Mythology so I decided to name my car Hesperides.

Hesperides -- according to Greek mythology, were the maidens who guarded the golden apples which Earth gave Hera on her marriage to Zeus. The sun is often in German and Lithuanian legends, described as the apple that hangs on the tree of the nightly heaven, while the dragon, the envious power, kept the light back from men till some beneficent power took it from him. Heracles is the hero who brings back the golden apples to mankind again. The apples appear to have been the symbol for love and fruitfulness.*

While I was in high school, I joined the National Guard, I could earn six dollars a month that way. That was so I could have enough money to take Helen to the movies. During the school year, we did our drilling on Thursday nights in the old Armory at Oregon Agricultural College in Corvallis. I was in the artillery division, where we were taught how to handle guns; assemble and disassemble them.

The summer after I graduated, I went to the National Guard's camp at an old army base near Jacksonville, Oregon for a couple of weeks. I rode the train from Corvallis down there to Medford. We practiced military maneuvers with machine guns and large artillery. Helen was real upset when I went off to camp. She didn't want me to go.

One of George's friends told a story about their high school days. Doris, who later married Bill, said; "One day, while George was still in high school, I saw him and the English teacher debating it out. He was standing there with tears in his eyes almost begging her to pass him." George needed to pass English so he could graduate from high school.

If she hadn't been good to me, I never would have gotten out of school. I didn't have much trouble with English I just didn't have anything to do with it! I didn't study much, wished I had. I never had a failing grade in school; I'd learn enough by going to class. I could always pass my exams pretty good. My daily work was worse than my exams.

---

* Encyclopedia Britannica, Eleventh Edition, Vol. 13.

George Shroyer at National Guard camp.

When Helen was in school she got good grades, got the best grades in her class. She studied all the time; she was pretty serious and she'd get after me because I didn't study.

As far as I can remember, the most trouble I got into in high school was when I pushed this kid out through the window. I finally paid nine dollars to get that window fixed. The funniest part about it was, that he was a year ahead of me in school and they wouldn't let him graduate without paying for the window. I loaned him the money so he could graduate. They gave him his diploma. I remember that it took all my money.

The next year, when I was ready to graduate, the school would not give me my diploma. They said I still owed for the window I'd broken. Turned out, that he never paid the money I gave him to the school. He told them I broke the window so I should pay for it. When Helen and I got married, the school still hadn't given me my diploma. We'd been married a year before Helen finally paid them the money and I got my diploma. That kid never paid me back, I had to pay for that window twice. I even grabbed him to keep him from falling out -- wish I'd let him fall!

Marie Shroyer went with George and Helen on their honeymoon.

I graduated in the spring and Helen hurried up and finished school at Christmas time so we could get married. I was nineteen and she was eighteen. We were just kids, but we thought we were pretty smart. We got married in Grandma Hatch's place. She lived there in Corvallis on the property that now belongs to Oregon State University. Rev. J. R. N. Bell lived in the house next door. Rev. Bell married us at about five o'clock that evening and Dad and Mama came to our wedding. Mr. Hunter wouldn't come. He was pretty upset and mad. He didn't like it that we got married. He told us; "You two made your own bed -- now you can lie in it."

We were married on Christmas Eve of 1926. After the wedding, I asked Rev. Bell, the preacher, how much I owed him. He said; "What ever I thought was right." So I gave him five dollars for marryin' us. We'd both been saving our money, Helen had seventy dollars, and now I only had seven dollars! Just about the same ratio as it is today!

Helen and I were going to live up there at Summit in the place where I was raised. We went back up to Summit that night. It was real muddy and it seems like it was snowing some. In the winter time, we had to chain up our pickup or car to get in. Lots of times even then; you couldn't get in. We had a shed that one of the farmers use to let us park our car in. That was on the Highway side, near Burnt Woods. We parked the car in the shed that night and walked in a mile and a quarter to the house. We came into Philomath for Christmas Day and took Marie back with us. She spent her Christmas vacation with us. After Marie grew up, we always teased her about going on our honeymoon with us.

CHAPTER 7

LIVING AT SUMMIT AND KINGS VALLEY

This was an exciting time for the young married couple. George's parents had recently moved into Philomath and Oliver agreed to let the newly weds move into the Shroyer Place, near Summit, therefore they had their own home during those early months.

Marie, George's little sister, was a frequent visitor during those early years. She and her mother were staying in Philomath that winter; while Oliver was traveling to the Holy Lands. Marie liked their Summit neighbor, Mr. Hymes who owned a small sawmill. A favorite memory of Marie's was of the times Mr. Hymes would give her a ride home from school. He often unloaded his lumber at the railway station in Summit in the afternoon. When driving the team, he stood on a small platform in front of the sled being pulled by the horses. Marie would stand behind him and hang on tight to keep from being splattered with mud or falling off.

Her parents always warned her to stay far away from the logging operations as logging was very dangerous. She recalls watching the horse logging operation at Cadwalder's mill with her father. Most people used horses for logging then and Marie said; "They were whopping big horses in those days"!

George started working for Mr. Hymes that first winter; handling the lumber and slabs while Hymes did the sawing. He quit after six weeks when Hymes couldn't pay him. Many mills were going broke and couldn't make their payrolls during those years.

> **I went to see Cadwalder about a job and he asked me what I could do; I told him I could do anything. He said; "Then you go over and run that planer. It sure was hard work." I lost a lot of weight that spring, rode Helen's horse Nig to work, and it was an eight-mile round-trip. I worked for him until we moved away; guess I did pretty good, he begged me to stay when we left.**
>
> **Helen didn't like it there; because the sun didn't shine much. We had our last big snow early in May and it frosted in there till June, you know. Helen planted a garden just before we left; so Dad and Marie could tend it that coming summer. She also planted a birch tree by the east side of the house.**

George knew Helen was right, about the limitations set by the weather in Summit. The gardens grown in that area rarely produced corn and tomatoes that ripened. The young couple decided they preferred to live at a lower elevation so they could have a good garden. Today, a stand of birch marks the rise where the old Shroyer place stood; silently attesting to Helen's love of plants and her nurturing spirit.

Horse logging in Kings Valley. Marie said, "They were whopping big horses in those days."

I went over and talked to Mac McCoy about working for him; he owned a sawmill in Kings Valley. He was glad to have me work for him again. He needed someone to load lumber and I'd done that before. He knew I was real responsible. We stayed with Mr. Hunter at the Hunter farm near Wren, for a few weeks and he was glad to have us back. McCoy had half-a-dozen shacks there at the mill. Whenever there was a vacant mill-shack the married men could move in there; Helen and I moved into one of those shacks soon as one was empty. The mill-shack we moved into was one-hundred yards from where the trucks dumped their loads of logs into the river.

Mr. Hunter was working for McCoy as the millwright. He kept the mill running, repaired and upgraded equipment. Occasionally he repaired cabins or built new ones. Helen drove him back and forth to work. After Bill and Doris were married they moved into the Hunter farm to take over the responsibilities there. Mr. Hunter moved in with us then and he lived with us most of the next twenty-five years, 'til he died.

Tin hat and caulked books. Tin hats did not become popular until the late 1960's.

McCoy's logging camp epitomized the "Roaring Twenties," an era of cars, planes, telegraph, telephones and electric lights. This generation believed—"They Had The World By The Tail." McCoy ran the biggest sawmill in the area, there in Kings Valley. Mr. Hunter built that steam powered mill. It had a brick dutch oven under the boiler to make it more efficient. McCoy had thirty-five men working for him and the mill cut thirty-to-forty-thousand board-feet of lumber each day. It was one of the largest producing mills around. He had several teams of horses he used for logging. Each team was a pair, with each horse weighing from one-thousand-five-hundred-to-two-thousand pounds. These horses were pulling heavy logs from one-quarter-to-one-half mile. The men that drove the teams made four-dollars-a-day. If they owned their own horses, they made eight-dollars-a-day. The horses would stay together in a big barn not far from the mill. The man hired to manage the barn was responsible for feeding, watering and bedding the horses for the night, as well as caring for any injured or ill horses. It was such hard work, a horse would only last from four-to-six years.

After the horses brought the logs to the rollway, McCoy's truck drivers hauled the logs on over the road to the sawmill. In those days, the average load consisted of five or six logs. Occasionally there would be a one-log load, containing twelve to fifteen-hundred-board-feet. A man named Walt Adams hauled a log containing eighteen-hundred-board-feet. That was the talk of the town for several years to come. Orders were for logs from sixteen to twenty-six feet long with a diameter of twelve to thirty inches. The most planks ran sixteen, twenty or twenty-four feet in length. They didn't take anything less than a foot in diameter and it had to be straight with no taper or limbs. The requirements were quite different from today's.

McCoy had a bunk house where the single guys stayed, he also had a cook house for the men that worked in the mill. Well, you pret'near had to, since most of them were single guys. They would come in and if you had good food they would stay awhile. If the food was bad, you'd have a hard time keeping a crew. Mill owners tried to get a good cook. That was the most important guy on the job other than the owner. If the men liked the cook and they had good lunches, they thought that was pretty good.

Starting to work for Mac McCoy again was pretty exciting. Helen and I'd gone to Corvallis and we'd bought me a new pair of cork shoes for my first morning at work. The cleats had pretty well worn off my old ones.

The Luckimute River ran right by the sawmill and it served in place of a mill pond. We opened the booms on weekends when the mill shut down. The first thing Monday morning, we'd close the booms on the river. This let us catch the first load of logs delivered that morning so we could float them down to the sawmill.

That first morning, McCoy asked Bill and I to help him close the booms. I was a little worried since I'd never learned to swim. Bill was standing on one boom and we were on the other and I was pushing down with my spike pole when it got stuck in the mud. I held on too long and fell in head-first, my cork shoes hung-on the boom. McCoy freed-up my shoes, reached in and pulled me out by my Mackinaw Coat. They said I looked just like a "Big Drowned Tom Cat." I had a whole head of dark black hair then. McCoy said; "I know what you were doing, you were just trying to keep your new cork shoes dry." That's how I started my first day.

Bill worked there as pond monkey. He was real good on the pond. His job was to bring the logs in to the mill. He was especially good on the logs out in the river. Falling in didn't bother him any as he could swim out if he fell in. He was a good swimmer. That old river would get up high in the winter time, driftwood, bark and stuff would build up. Old Bill would go out there, open the booms and let that drift wood go on down. He watched the pond pretty good. He could hardly walk on the ground because he was pigeon toed; on the logs he seemed to stay up real good.

It wasn't long before I became boss of loading the train cars. I was just twenty years old, and the youngest man. I was responsible for the bill of lading on each train car and it told the train crew where to send the lumber.

(LEFT) George, wearing his caulked boots and with dinner bucket in hand, ready for work. (ABOVE) George and Helen's mill-shack in Kings Valley.

Marie, warmly welcomed by George and Helen, visited often. She recalls setting on the front porch of the mill shack watching the trucks unload into the river only one-hundred yards away. There were great splashes as the logs dropped into the river. Several trucks unloaded each hour of the day.

Some of Marie's fondest memories include picking little wild strawberries with Helen. The berries made a deep red jam. Mid-to-late-summer brought their search for rare, small wild black berries. These grew in logged over areas and made wonderful jam also. Late afternoons on hot days they hiked a quarter of a mile up the Luckimute River for a refreshing swim. That was certainly a treat. When fall arrived, their activities turned to canning fruits and vegetables. Marie enjoyed helping to make applebutter over a hot stove as she believed nothing smelled better.

The exuberance of youth and the struggle of life filled the mill. This was a bustling community with few comforts and many dangers. In the evening the men took their turn swimming in the river. Quiet laughter filled the evening air as the young people shared their day's events. Helen often worked on her embroidery as they sat out on the porch in the evenings quietly watching the ever lengthening shadows bringing with them cool breezes from the coast.

Logging in those days was one of the most dangerous occupations. There was an excitement shared by the men as they braved the dangers; these life threatening experiences bound them together. The size of the trees contributed to this as did newly designed equipment, not yet safety tested, unskilled workers and weak materials. This was an era of unprecedented development in the timber industry.

Helen said:

I remember that George fenced an area around the shack in which I grew flowers. He used chicken wire for fencing because the neighbors had chickens. We never had a vegetable garden in Kings Valley. When we needed just a few things I'd walk to the general store in Kings Valley to buy groceries. Alice Albertson ran the big country store. I always prepared all of our meals at home – we never ate at the cook house. I would have the meals ready for George and Papa.

George and Papa needed quite a few clothes. Their jobs were dirty ones. The washing took me quite some time with the old fashioned washboard and tub I used in those days. I would go into Corvallis to buy work clothes. We all liked to read, even Papa. I'd always stop at the Corvallis Public Library and get books for us to read each time I went in. Katherine LaBare was our neighbor at the Hunter farm; she would take me with her to the Corvallis Public Library when I was in high school.

George said he remembered when he and Helen would go to Corvallis to see a movie on Saturday nights and would drive into town in their 1923 Model-T Ford Roadster. It had no side windows and it was often cold so he would "put a blanket around Helen and we would set real close."

After the first winter at McCoy's mill, I used some of our savings to buy Helen a car. It was a Dodge coupe. She needed a car that she could drive. You could break your wrist cranking those Model-T's. I didn't want that to happen to Helen.

In those days Helen and I, we'd split the check both ways; she would take half of it and I'd take half of it. We were going to see who could save the most. She would pay for the groceries and the up keep of the house with her half. I would take care of the car, gas, movies and candy with my half. I remember I could get six bars of candy for a quarter at the store. Once or twice a week we'd go in to buy candy and she'd say what kind she wanted. When she was buying groceries and stuff, I don't remember her ever buying candy.

One time I was out in McCoy's shop talking to Mr. Hunter when this guy came around. He wanted to sell me one of those mandolins, I sure liked to hear those instruments. He couldn't play but he got this old blacksmith to play. That guy could play it by ear, he just picked that thing up and within a few minutes he was playing tunes on it, he was really good. Good Lord! I thought if that old guy could play a tune that quick; it would be no trouble for me. So, I bought one for fifteen dollars. That was quite a lot of money in those times and it was about one-third my half for that month. I thought Helen would learn to play it as she didn't have much to do, no kids or anything.

I took it home to show Helen. Boy! She wouldn't have nothin' to do with it. When she found out I bought that; she jumped me for throwing my money away. I told her, I thought it was enough that she was saving money. Helen said; 'If you're not going to save money I won't either.' Well! She just grabbed up her money and threw it right in that stove, that was fifty-five dollars. I was there, so I just grabbed 'er right out. Ha! Ha! Ha! Boy! Helen got real mad. She pretty well took over all the money after that. She paid the bills and did a good job. I was more in the habit of trying to spend some of it to have a little fun.

Helen's firm belief that saving money would provide you with security and independence came from her high school economics class. She had been taught that to get ahead in this world you would have to save part of your income. This, combined with her personal strength, provided the potential for their future financial success.

I remember I wanted to get on trucking. Old Mac was canning someone almost every day. He had about five or six guy's hauling in there with company trucks. I remember I asked him one day; How 'bout me getting to drive truck? He looked me right in the eye and said; 'George, you couldn't even run a wheel barrow.' So I thought I'm never going to get to drive a truck here.

Two or three mornings later why, he comes over to our place. I heard somebody hammering on the door before Helen and I got up. We didn't have to get up very early because we lived within five-hundred feet of the mill. He told me I could take "Old Long Tom" out. Everybody knew the trucks; they had different names painted on them. This truck had been lengthened out, about three feet longer than the others. I knew which one it was, so I just got up and went to work. I didn't eat any breakfast. The old woods-foreman asked me if I knew where to go. I told him yes. I thought I knew where to go cause I'd gone by and seen 'em hauling out of there.

When I got out there in the woods, nobody was there. I went to loading this load of logs by myself, it took me an hour and a-half or two hours before I got back to the dock to unload. Old Bud Reynolds, the woods-foreman, jumped me; he wanted to know where I'd been. I told him where I'd gone to get the load of logs. He said; "How did you load them?" I told him with a peavy. He said; "Well, we've been using a team with a cross-haul on that one. Ha! Ha! Ha!"

They had been loading these logs with horses; I'd gotten' them on by myself. That was pretty hard to do. This rollway was pretty steep. So, when I loaded those logs with the peavy, I'd stick my foot under it to hold it from rolling back. I did that, but then it got to smashing my feet so bad that I had to give that up. I got me a rock and the next time I rolled the log up about six inches or a foot, why I'd kick that rock under the log to hold it from rolling back. I'd used the peavy quite a lot to load timbers so I knew how to use a peavy.

It kinda tickled the old woods foreman who said he would go 'right back with you and get you to the right place' he said. When Reynolds went back with me, he took me to the place that I was supposed to have gone to. The one he took me to was built to be loaded the regular way, with a peavy. It slanted towards the truck instead of away from it. I'm sure McCoy got a laugh out of that when he heard about it from Reynolds.

We didn't have any dozers in those days. We loaded the trucks with a team of horses or by hand with a peavy. If you were building a rollway for loading with a team of horses, you'd leave the rollway pretty steep. It would take quite a lot of pressure to get the logs off the rollway. They made them steep. When we built rollways, we had to dig them out by hand.

You were building a rollway for loading the logs by hand with a peavy; you'd build something they called a one or two-drop rollway. When the logs came in from the woods they were pulled up on the rollway, then loaded on a truck. There was a crew of men there to help load the trucks. The lowest drop was even with the bunks on your truck. They loaded those two or three logs whatever it would take on the bunk. There would be another one about thirty-inches higher; then they would put a plank on the next rollway up. They'd have to work from there over to your truck, they would roll the logs on. If two or three people put a load on, it would be pretty easy, one guy on a rollway wasn't enough.

Not long after I started driving truck; McCoy told me to take this Model-T Ford truck. It didn't have a door on the driver's side; it didn't even have a roof on it, the old canvas top had been torn off. All it had was just a windshield and a box over the gas tank, so I fixed up a burlap bag stuffed with sawdust to set on; I drove that for some time. This truck had a little half moon key on the steering wheel to hold it on while you were drivin'. After a while, with me jerkin' on the steering wheel, that half moon key loosened up. I'd never had the half moon key off, it'd been stuck on there; now, I found that I could take that steering wheel off. I'd meet the other truckers coming back, or some old farmer and I'd scare 'em to death by showing 'em my old steering wheel. Good Lord! Those guys thought that was it—they'd give me the whole road, you know. Ha! Ha! I'd put it back and go right on. A truck as a rule, if its wheels are lined up, will stay lined up, for a little while.

This one time, I was coming down through some rolling hills where there was this little dip; it had lots of chuck holes. The half-moon key fell off and the steering wheel came off in my hand; it sure didn't worry me because I knew it would stick right back on there. I stuck it back on but the darn thing just spun around. That little half moon key wasn't on the floor board and of course I couldn't look around; had to look ahead and run the truck into the bank. The logs came up and took everything right off to the motor; I went right-out over the windshield, just ahead of the logs.

Old Art Lyday lived just across the road from there; he came over wantin' to know if I was dead. I told him I didn't know, but I could walk so I guessed I wasn't but it kinda beat me up all right. Then I went in and I thought I'd get canned that first month I'd been on the job. When I told McCoy, he said; "Well, what are you doing about it? Get it in and get it fixed." McCoy was real fair; he'd give you a chance.

It broke a couple of lines to the gas tank, and the windshield and the truck had to have a new reach to the trailer. Mr. Hunter helped me put that in. We used an eight-by-eight-inch coupling-pole twenty-feet long. We got that truck going again, right away, probably the next day. We didn't have everything on it. That time, when the truck came out of the shop, the steering wheel wouldn't come off. They'd fixed it permanently and that sure relieved lots of farmers around there. They didn't like to see a truck to begin with; then, to see one with the steerin' wheel off it. Ha! Ha! I'd been doing that for a while and I'm sure McCoy knew about it.

That wasn't the only trucking accident I had. I had different accidents all right, but that could have been a serious one; that could have killed me. Those were good-sized logs I had on there; they were second growth. I'd say there were about four-or-five logs on there. It was sure hard work back then, pretty dangerous too. I've been awfully lucky; I don't know why.

I broke my wrist cranking that Model-T Ford truck; I could drive with one hand but I couldn't crank the truck. McCoy wanted me to be drivin' so he bought a four-cylinder Chevrolet Truck with a starter, It started real easy. It was the only one McCoy owned. Most trucks were four-cylinder until 1929; that's when they came out with six-cylinder engines. That same year Ford came out with their Model A's.

Weren't many winter jobs in those days. McCoy kept only his best men to drive his Nash Quads. There were only two men hired to drive during the winter; those jobs went to the two guys that drove their trucks with the least repairs and hauled the most logs.

Pre-1920 heavy duty utility trucks. (ABOVE) **The Nash Quad was a 4-wheel drive truck that was popular in the First World War, with fleets of them sold as surplus after the war.** (—Bert Webber collection) (LOWER) **Unidentified truck. Note solid rubber tires and leather bench seats in both trucks.**

(—Benton County Historical Society)

Helen and I thought I ought to have that winter job. We hadn't been married long and we needed the money to make it through the winter. Helen and I had a code so I could let her know when I would stop to get my dinner. We were hauling on the dirt in the summer time and we dumped our logs into the pond about one-hundred feet from our shack. I would hold out one finger, and make one more trip; that would be about an hour. If I held up two fingers why, I'd make two trips. If I gave her a zero, I was coming in right then for dinner; so she'd get dinner on the table. We saved time by Helen putting down magazines for me to walk on so I didn't have to take my cork shoes off when I went inside, you know. I'd get to eating soon as I came in and I'd get right out; this was so I could get ahead of all the rest of them.

McCoy owned a couple of these trucks, from World War I; they were four-wheel drive; so they called them quads. The two trucks were one-unit, with about forty feet between the cabs; they were setting back-to-back with a coupling pole. Just like a train today, with one engine on each end. You drove it like a regular truck; you did the shifting, ran the gas pedal and all of that; only thing is; you didn't do any steering.

They ran on wooden rails, on a road that went out a couple of miles from the mill. The men built these rails in the fall just before the winter rains set in. Sometimes they built the rails out of eight to ten-inch diameter trees, real nice ones; they'd smooth them on the top and on the inside. Other times, they used old two-inch by eight-inch boards stacked up right on top of each other. They had hard-rubber tires with flanges on them so they wouldn't fall off the rails. The Nash-Quad with the short wheel base was the one going into the woods, empty. Wherever it was steep, the guy behind would put his truck in reverse and help by pushing. I drove the one going into the mill loaded with logs; I did that for two winters.

I worked with Hurley Morgan the first winter. He was a tall lanky guy and I liked working with him. We had to load the logs in the woods using rollways. We'd bring a load out of the woods and dump it into the river and the logs would float down river, about six or eight-hundred feet, to the mill.

When we were unloading, the man closest to the pond had to jump over those logs or stand out of the way. Only thing for you to stand on was a log and if you weren't good at it, you might fall six to ten feet. Morgan was real good about getting out of the way. Next winter Morgan went to work for somebody else and Ed Albertson got that job. I liked working with him. He could tell stories from daylight to dark and never tell the same one twice. Ed was older, but he wasn't married. He'd try to get us off the track by pulling real hard when we were heading for the woods and cause the wheels to drop off the rails down into the mud. He'd get us off, then he'd set around and laugh. I'd make a short railroad to run it back onto the track, then I'd have to jack up one of those Nash Quads to get it started back on there. Sometimes it would take an hour. I was furious; I wanted to be hauling logs.

Our job was to keep enough logs down there for the mill to run on. Ed was just putting his time in. It was all just a big joke to him. One time Old McCoy happened to be coming over there and he saw him doing that. I remember McCoy came over there and he canned Old Ed. He said; 'Now Ed, if there is anything that I haven't given you a chance to do, you let me know. I want to be fair with you. Everything I put you to do, you can't do, you always fowl up.' I remember seeing Ed going; seeing him laughin' out there, with those big old gold teeth he had in front.

Later, Ed got to having his own sawmill where he did real good. He just wasn't grown up yet when he was working for McCoy. He was probably just twenty-two or three; wasn't much older than I was, two or three years. When he grew up, he quit all his joking and settled down to business.

McCoy sent me to the Simpson Brothers Sawmill to haul out some lumber. The Simpson Brothers had just started up this mill with about twelve to fifteen men working there. They lost their father in a logging accident a couple of years earlier. Mrs. Simpson, their mother, cooked for the loggers there. On a quite windy day she stepped outside and walked just a little way from the cook shack to call the loggers in to dinner. As she did, a great gust of wind hit the camp causing a large tree to fall on her killing her in an instant. That tall second growth tree was split in half by the wind. This was the first person I saw killed. She was only in her fifties. It was really hard on me but I helped her sons carry her out.

Another time there in Kings Valley, I saw a bunch of guy's I knew, that were practicing their high climbing for the 4th of July doings that was coming up. They were getting ready to compete for some prize money. Orville Edwards, his brother, and several others were there, climbing a tree that had been topped. That tree was only about eighty-feet tall. When one of those guy's was up near the top, he wasn't paying enough attention and threw his climbing rope up over the top of the tree. That caused him to fall over backwards. He fell to a hard plank area below killing him instantly.

These deaths heightened George's awareness of danger and crushed any false sense of security he might have had. He realized that he could have lost his own life a few months earlier when the steering went out of his Model-T Ford

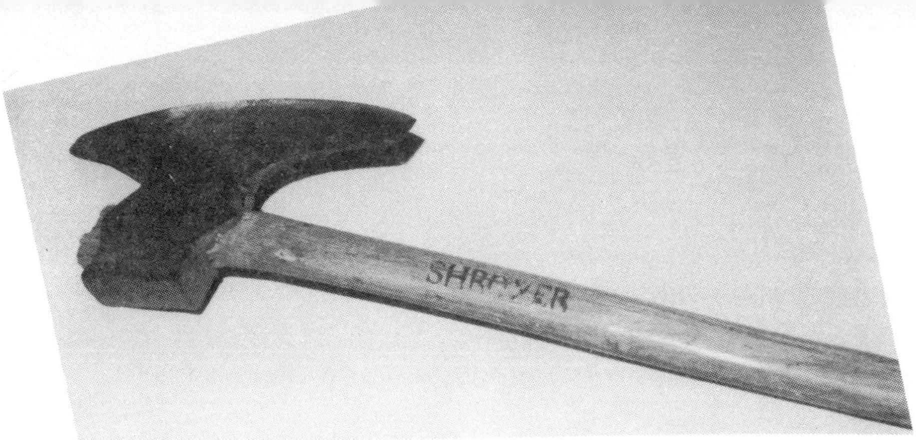

George Shroyer's 200 year old broad ax.

truck. The loss of life in this profession was becoming a hard reality for George.

That same year we had a summer that we called the summer with no rain as it had been real hot and dry. By fall, fires were breaking out in the woods and at the mills. The first fire McCoy's crew fought was at a mill close by. Their timber and slash caught on fire and it was threatening another sawmill. It took fifty men working for three days to put out that fire, but we saved the mill. This tall lanky guy, by the name of Story, worked for McCoy. He was helping us keep water on the mill to keep the cinders from catching the roof on fire. I put up with him all day telling me what to do. I waited until he was close to the pond and I let him have it with the water hose. He fell right in – *swoosh*! Everyone thought that was pretty funny; he'd been telling everyone else what to do all day long but he never told me what-to-do after that.

Even though it was early winter, the rains hadn't come and it was late November of 1928. Helen and I woke up one night and we could smell smoke. A fire started during the night; no one ever figured out how it started. Some of us thought a spark had gotten loose from the open burner and started in the sawdust under the mill. The mill was tinder-box dry and the fire burned real hot. We all tried our best to save the mill that night. We saved a lot of the equipment but we couldn't do much more. Mr. Hunter hurt himself fighting the fire but he didn't notice he'd sprained his shoulder until later. Our shacks hadn't burned, but the mill had, so after the fire all of us moved out as soon as we could find places. Helen and her father and I moved over to Gellatly Canyon.

When McCoy's mill burned down in Kings Valley, the river was full of logs and they couldn't saw them because the mill had burned down. There were four of us hauling equipment and logs out of there. There was Walt Adams, Hugh Pleasant and myself, and a guy called Hamilton. We hauled the logs down to McCormick's sawmill near Wren. We helped McCoy move the equipment we'd saved from the fire. We took this over to the Soap Creek area where he was building a new mill.

George and Helen Shroyer, a happy, young married couple. George shown at the Hunter farm. Helen, who loved to raise both vegetables and flowers, in her garden in Gellatly Canyon.

CHAPTER 8

LIVING IN GELLATLY CANYON

Gellatly Canyon is located a few miles west of Wren on the highway to Newport. It is named after an old pioneer family that owned three homesteads in the area. This was Helen and George's first home away from the mill. There they were nestled in the canyon. This seemed more like a real home. The house set above the highway, and it had wooden stairs leading up to it.

> **Helen said:**
>
> **We were glad for the stairs since it snowed quite a bit that December and January while we lived there. It looked beautiful with the pretty curtains hanging in the windows. I called this the "Doll House" because it was quite small. It had one bedroom, a kitchen and a living room. It did not have a porch.**
>
> **George said Helen decorated the "Doll House." She hung the muslin curtains with pink ruffles around them, on the windows. She'd made these while we lived at Kings Valley. She also hung pictures of horses and the family on the wall. She had lots of house plants that she had started. We made this little place look real home like.**

Helen made dish towels from flour sacks and she embroidered flower designs on them and blanket stitched the edgings. The salt sacks made good dish clothes. They lived in the "Doll House" for a couple of months.

> **I'd asked Old Walt Kisor if we could rent the mill-shack he owned in Gellatly Canyon. It was pretty close, about two or three blocks. He said that we could, soon as the family moved out of there. We'd have to fix it up some; we all worked hard cleaning it up after they moved. They'd used rough cut lumber for siding. The boards ran straight up and down; they had cracks between them, the inch-wide cracks had one by four's nailed over them. They called this batten and board siding; lots of mill-shacks were built like that. I put tar paper on the floor to cover up the half-inch cracks there, and on the inside walls where the wind would come in; so we could keep it warm in there. We had, a cooking stove and a heating stove, really got it fairly warm with the two stoves.**
>
> **Back then most people were having kids right away. Helen and I were married for about five years before the girls were born. People would ask why we weren't having any kids. I use to say; "What in the devil! The kids were falling down through the cracks in the floor and we couldn't get them out." Sometimes I'd tell them Helen was bashful! After the girls were born I use to tell them they'd have more brothers and sisters; but the others weren't strong enough to crawl back**

(TOP) **The "Doll House" in winter.** (LOWER LEFT) **George and Mr. Hunter in Helen's flower garden.** (RIGHT) **George bought a 1926 Dodge Coupe for Helen.**

through the cracks. Well! Those houses were kinda drafty, had lots of cracks in them; but not quite big enough to lose kids in.

Helen put her curtains up again and fastened them with string and nails, they fit the windows real good. The bathroom was an outside two-holer. There wasn't any running water in that old place at first. Mr. Hunter and I cleaned out a little spring up on the hill and run the water over to the shack. We took our bath's near the kitchen stove in a large wash tub.

Mr. Hunter had built the shack for McCoy and his wife while they ran the mill there in Gellatly Canyon. It was larger and nicer than most mill-shacks. It had a porch across the front of it, a large kitchen and a store room; a large living room and two large bedrooms. We moved in there, February of 1929; just in time for Helen to plant her peas. Helen, Mr. Hunter and I wanted to raise a big garden. We thought we'd do pretty good, with a big garden. There wasn't an orchard in there but Helen planted some fruit and nut trees later.

Helen and her father put up lots of food that year, about five-hundred quarts of fruits and vegetables. That was the beginning of the depression. Helen's father never worked out again after the mill burnt. We had a wood shed outside, a small chicken house, and a field for the livestock. We soon got a milk cow, some chickens, and a bull. With her father's help, Helen tended to the animals. He generally milked the cow before breakfast and gathered the eggs before he started his work. This would be our home for the next few years. We paid Mr. Kisor five dollars a month for rent. This would be the only time in our life that we ever paid rent other than the year we moved to the coast in 1942. We always owned our home after we moved out of the old mill shack there in Gellatly Canyon.

When we moved to Gellatly Canyon, we had two cars, my 1923 Model-T Ford Roadster I'd bought while I was in high school and our 1926 Dodge, I bought that for Helen in 1928. After that first winter at McCoy's Mill, I used some of our savings to buy Helen that Dodge. I bought her one she could drive, it had a starter on it as well as a crank. When cars first came out, the only way you could start one was to crank it up. My Model-T Roadster had to be cranked, that was the only way to start that one. I hurt my wrist several times cranking it, the crank on it was dangerous. If you were not strong and didn't know "Just How To Do It" you could brake your arm or your wrist trying to start it. I wouldn't let Helen touch that one, I didn't want anything to happen to her. I'd taught her to drive while we were still in high school. I always thought every woman ought to know how to drive.

Helen could start the Dodge with a key unless the battery was low. She would take the Dodge when she went shopping and to the library in Corvallis. She would pick up Marie when she came through Philomath. Helen would bring Marie out to stay with us for a few days.

Helen said; George and I wanted to save all of the money we could; we didn't buy a new battery until we absolutely had to. We didn't want to waste any of his hard earned money. Once we moved to Gellatly Canyon I went to town more often. Sometimes Marie and I found that we would have to try to crank the Dodge, we'd walk down the highway a couple of blocks or more to ask Mr. Beech or Mr. Ross to come up and crank it for us. Once they started it we could drive to town; the battery would recharge itself as we drove. It normally started with the key after we finished our shopping.

Mr. Bill Ross had a daughter-in-law, Hazel. I use to take her with me sometimes when I went shopping, the two of us would go in to town together. We would stop at the bank to cash the checks, then we bought our groceries. I would save out enough for other things and put the rest in our savings account at the bank; I saved the greater part of it. Almost everyone else was having trouble just getting by and we were saving money every month. That's how we were able to build our new house in Gellatly Canyon.

We got by on less money than others because I raised such a large garden and I always tried to can at least five-hundred quarts of fruits and vegetables each year. George and Papa would wait until early winter to help me can the beef. We tried to can at least fifty quarts of meat each year.

Helen's greatest contribution to the family's financial stability was her large and productive garden. She started planting early in February each year. First, she put in early peas. Next she started seed for one-hundred tomato plants. She always had a good garden out there in Gellatly Canyon. She started with early peas, onions, radishes and lettuce. Then the regular garden included all the vegetables, corn and tomatoes. They ripened just fine in this location. The weather was warmer and the seasons longer than they were in Summit which was at a slightly higher elevation. Summit wasn't too good for corn and tomatoes because it frosted fairly early in the fall. Helen knew which variety of corn to plant to make sure it produced in the canyon. She always planted "Early Bantam" corn. This was the fastest maturing corn she could find. It didn't grow very tall, maybe four to four-and-one-half-feet tall but it always ripened and produced small, light golden ears of tender corn. She started her tomato plants early and kept them well mulched. She covered the young plants during cold spells. She always planned on putting up at least sixty to eighty quarts of tomatoes.

(LEFT) **Lucy and Joyce sitting on the porch steps of their mill shack.**
(RIGHT) **The Shroyer family. Helen is holding Lucy. George is holding Joyce**

We always had a good garden; Helen did most of it. Mr. Hunter and I always helped plant the garden in the spring. I ditched water in from a little creek above so we had water running in there for the garden. We'd run it down the ditch, then we'd open a different little ditch each evening so it would run on a different place each night. We had a good little irrigation system there. That garden would grow real good in the summer time with all that water.

Helen and her father were real generous, those were pretty difficult times back there in the Thirties. We never went hungry but lots of people were having a pretty bad time. Some old men lived in shacks close by, they were Old Man Beech and Old Man Ross. Mr. Hunter was always carrying food to them. He also took food over to Bill and Doris and their five kids.

When I run the cut-off saw I got the first pick of the wood so I'd bring the wood home from the Rat's Nest sawmill where I worked. That was nice; it would be in small enough pieces already. If it was longer than you needed; well, all you had to do is to cut it up with a saw right there. I would bring a whole pick-up load home every night. That gave us enough wood for winter 'cause, one load would be enough to last a week. It's quite a little job to put it in your pickup and unload it and pile it up when you got home. Mr. Hunter always helped me unload it. We'd have to carry it a hundred feet or so to get it placed in our wood shed. Eventually I got it fixed so I could drive in there close, couldn't do that at first.

I'd get enough wood that way to do us all next winter. My Dad, taught me to do that. This was done before it was time to put the garden in each spring. We always had our wood cut up very early before April then all summer it would be drying out. In the fall, we took the wood in as soon as it began to rain again. In October we'd split it and stack it in the shed. That way you would have dry wood to burn the following winter. Mr. Hunter liked that wood idea, bringing it home eight months early. He'd always cut his wood in the winter time, he'd never had any wood cut up ahead of time, never. Yes, he liked that, he liked having our wood in early each year.

I drove past McCormick's mill on my way to work each day. At that time, when I'd bring my pickup home at night; I'd have it loaded with wood; tie ends and three by twelve's and stuff. That dog would always come out, run me up the road and bark at my left front tire. One night when I was coming home, I got me a chunk of wood that had whiskers on it. You could hold it by the whiskers with one hand; I was going to teach that dog something. That was the reason I put that piece of wood down there by my feet. I got down the road a bit and this dog came a running out there, right after me, a biting at my tires. I just held this chunk of wood out and hit him right on the head. Boy! He sure got his old mouth full of gravel. I remember, next morning when I came to work he heard me coming and he started to come out there. He saw who I was -- so he didn't come out after me. I saw Old McCormick a week or two later. He said; "You sure cured my dog of chasing you the other evening." He'd seen me doin' that, I didn't know he was watching me. I'm glad it didn't kill his dog, that it only cured him. I suppose he chased all the other cars, I don't know but he sure quit chasing mine.

George and Helen were just starting their life together. George's humor began to develop as the hard times worsened. They certainly worked hard in those days but they also had a lot of fun along the way. Money became scarce as the depression deepened and they worked courageously to struggle through the challenges they faced.

George had confidence in his ability to do anything. He also had confidence in Helen's ability to pursue opportunities that would insure their economic independence.

He encouraged her full participation in developing their strong economic base. George considered Helen's personal strength a desirable quality. He did not

Helen, in 1929, standing by their service station they built along the highway in Gellatly Canyon. Price of gasoline was 23¢. Helen's Dodge car in background.

try to stymie her growth as an individual; instead, encouraging and assisting her whenever possible.

We'd been talking about buying land, Helen wanted us to own forty acres. So after the fire we bought forty acres there in Gellatly Canyon. It was across the highway from the old mill-shack we were renting there. We bought that from Old Walt; land didn't cost much in the depression years. Later we also bought the land that the mill-shack sat on. It was a few years before we could do that. We wanted to build us a place to live like Helen wanted, we built that later. It lay on the south side of the highway that goes to Newport.

We decided that Mr. Hunter and I should build a service station and Helen would run it. We were going to try to make more money, we figured we'd sell quite a little gas there. We put that service station in right away, in about 1929. Mr. Hunter built most of the service station. I ran a pipe under the road to provide the service station with drinking water. We tried to put in electricity at the same time. We put in a little diversion dam on the creek there on our forty acres. Then we put in an eight-hundred-foot flume and a water wheel. He built the wheel and I put the generator on it; I set up the wiring and stuff. That first attempt wasn't too safe, wasn't too dependable.

George and Helen's unfinished house near service station. Note gravel highway. They would have to wait another year to finished the house.

We'd pretty much finished building the service station but the land around it was soft and muddy. We wanted to make it easy for people to pull up to it, we didn't want people getting stuck there. Back then when you wanted to make a parking place; you made it out of different things than they do today, some were made out of old planks or wood.

There was a gravel pit up above us there in Gellatly Canyon, nobody could afford to buy crushed rock back then; even the county couldn't afford to buy very much of it. Ha! Ha! Ha! The state bought quite a little of it. People didn't have access to gravel like they do now and we sure couldn't afford gravel. We got to thinking that maybe those cinders from the old steam powered sawmill might work. They'd had a big burner in there; probably about eight-feet-by-eight-feet and about twelve-feet high. They used it to burn the bark and other waste wood from the sawmill. That often had mud on it; when the mud was burned, it would create these cinders. They were kinda like brick, only they were in little pieces. We figured that would make as good a road as anything. So, Mr. Hunter and I were going over to that old sawmill and cleaning out those cinders. We'd haul them over to our place and throw them on by hand.

We got turned in for hauling those cinders for our service station. We were hauling mostly on our own road with our little pickup. We were trying to save money so we didn't have a license for the pickup. We were only hauling across the main highway for about five-hundred feet. Somebody turned us in to the cops and they took our last ten dollars. We'd gotten down to our last money, our last ten-dollar bill!

After George and Mr. Hunter finished the gas station Helen ran it. Those first three or four years required that Helen constantly run down the hill and across the highway to mind the service station when people stopped for gas. She virtually ran up and down a couple of hundred feet of plank sidewalks. Their shack, in Gellatly Canyon, set on a hillside at about a seven to ten percent grade above the highway. She kept the books for their business as well. Her accounting was accurate to the penny.

Helen did far more than most young married women. Tended to her large garden and canned their year's supply of fruits and vegetables and looked after the livestock. She cooked, cleaned, washed for herself, her husband and her father and planned for their new house. All of this and tended the service station. During this time, she went through two pregnancy's giving George two beautiful young daughters. She also went through a desperate illness nearly loosing her life. Soon after Joyce and Lucy were born, George's endearing term for Helen became "Mama." George also began referring to Helen's father as "Grandpa."

Mama and I had two babies right there in the middle of the depression. Joyce was born in 1931 and Lucy the next year. We had the money to pay for the hospital bill when Joyce was born, we never charged up any bills like that. We always knew ahead of time that we was going to have some medical bills. We would do without other things and save our money. We didn't go to shows, drink, smoke or anything else. If we had, we wouldn't have had enough money to buy the forty acres, build the service station and the new house. We wouldn't have been able to buy any orange juice and cod liver oil for the babies, that was pretty expensive.

Mama and I decided that we could save money if she had Lucy at home. We wanted to save as much money as we could. That was right in the middle of the depression and a month's pay was only twenty-eight dollars. Later, when the bill came, we found out we would have saved money if Mama had gone to the hospital. Dr. Whitelaw charged us sixty-five dollars to come out to our house when she was born, this was more than two month's pay!

> Caring for babies was pretty hard in those days. Mama was doing all the washing and diapers on a scrub board until about half-a-year after Lucy was born. We decided Mama needed an electric washing machine so we bought a new Delco generator. When we bought that, we made payments on it, that's the first time we ever made payments on anything. Grandpa and I installed that. Once we got that in there, why then we were able to get a brand new Montgomery Wards washing machine. It had been pretty rough living before that, we had it as long as we were there.

Later, Helen had a serious bout of appendicitis. Since childhood she had suffered with it periodically. One of Joyce's earliest memories is of her mother's critical illness. Helen always said; "I made it through that experience because I knew George couldn't take care of the girls by himself." Helen had tremendous amounts of energy and determination. Joyce still remembers hearing her mother crying-out in pain.

Helen's father became frightened by her incessant pain and ran out the door; saying he was going to get Bill to take her to the hospital. Helen's father, was seventy years old, he ran one and a-half miles over the hill to the Hunter farm. Joyce and Lucy were left alone with their Mother who was writhing in agony. The two young girls, only two and a-half and three and a-half years old could do nothing but watch as their Mother cried in pain. It left an indelible mark in Joyce's memory. The time that it took for Grandpa to return with Bill must have seemed like an eternity.

Joyce recalls watching her Uncle Bill as he wrapped her mother in a blanket and carried her out to his open car. She could not understand where he was taking her mother. Helen knew Bill had come to take her to the hospital. She cried out that she wanted to wait for George to come home. Bill said; "Sis – you have to go now." He drove off down the road leaving Joyce, Lucy and Mr. Hunter to wonder if they would ever see her again. Helen didn't even know anyone having had an operation and the idea of having one was a frightening prospect. Her father sat in the rocking chair, crying after Bill left with Helen.

Much later, Bill returned looking very grave. Joyce and Lucy went over to stay with their Uncle Bill and Aunt Doris. What a busy place they had with four boisterous youngsters of their own and two very young nieces; also, Doris was pregnant again. George had been working at the Rat's Nest when Helen became so ill. When he arrived at the hospital she was in the operating room. They were in there several hours. After they came out, the doctors told him that Helen's appendix had ruptured some time ago. Peritonitis had set in and she was critically ill. Now, only time would tell whether she was strong enough to fight off the infection.

Helen was desperately ill during the next couple of weeks. There were no antibiotics during those years and no one was sure if she would make it. George recalls trying not to think about what he would do if Helen died. He worked hard and kept busy, that was George's way of blocking out the terrible possibilities.

In the evenings during those first weeks, George would come over to see his children. Joyce recalls the guarded, solemn looks that the adults exchanged. The adults kept shaking their heads and talking about her mother not being, "Out Of The Woods" yet. Joyce said; "I couldn't imagine why Mama would go into the woods, it didn't seem like a very good place to be." She often wondered if she would ever see her mother again.

It was several weeks before Helen could return home, Dr. Whitelaw would not let her take care of the girls for at least another month. Helen's father cared for her during her convalescence and the girls stayed at Bill and Doris'. Helen had been intensely ill and this put a strain on all of them. The money they had been saving for the new house went to pay medical bills. They would have to wait an extra year to finish the house they had started to build.

> Old Man Beech and Old Man Ross lived there in Gellatly Canyon not far from our place all those years. Old Man Beech was a lot of help to Mama when she was able to take care of the girls again. He would sit out in the yard with the girls while Mama worked in her flowers or in her garden. He used to play with the girls a lot, sit on the grass and talk to them. Joyce and Lucy were always happy to see him. Old Man Ross, he wasn't so much that way. He was a good neighbor all right, he would help Mama whenever she asked.

The winter Helen was pregnant with Joyce, she started thinking about what kind of house they could build. One day while reading their newspaper, the *Oregon Journal,* she found an advertisement for a house. It had a picture of a "Thirties Bungalow" style house. An article describing the house indicated the house plans cost ten dollars; Helen sent for the plans right away. George and Grandpa thought those plans would make a fine home. Grandpa built most of it by himself as time and money permitted. As soon as they saved some money ahead, Grandpa would build some on their new home. They continued to live in the old mill-shack from 1929 to 1935. George was working long hours and away for several days at a time. This took a few years as times were hard. It was 1935 before they were finally able to move into their long awaited "Thirties Bungalow."

> It sure was nice having that new house when we got through building it we moved right in, we were real happy to be in there.
>
> I was working in my shop repairing equipment for Rex Clemens when a stove salesman came by. He was selling those "Home Comfort Stoves." They were one of the best stoves you could buy, but we

The "Home Comfort" cast-iron stove.

couldn't afford one. He was bound on selling us one. Well, he's telling me how the lids would never warp, or break, or anything. Finally I told him; "If I couldn't break one; well, then I'd buy one." If it would break, I was to get one free. I was sure it would break; so I took a lid and put it between two cat-rollers; it was up off the floor about eight inches. I got this big twelve-pound hammer with a splitting edge on one side of it and I hit that grate right in the center of it. I was supposed to be able to hit it twice. I hit it -- it bent but never broke; so we ended up with a new stove on me!

It was real nice looking, made of cast iron, about a-half-inch thick. This was covered with a dappled gray enamel. It had a nice cooking oven and there were two warming ovens up above the stove. Several drying rods allowed towels and clothes to dry about twelve inches from the sides and back of the stove. You could set on the oven it was so strong; it was an awfully good stove. Mama didn't like it at first, because it would take a couple of loads of kindling to get it hot. It was a heavier stove, didn't heat up as quick as the ones we had before, but it kept the heat better. Soon as she got use to it, Mama liked it better, you know.

I don't remember how much it was, but we've still got it. It's in the basement today. There were big reservoirs for warming water, one on each side of the stove. It wasn't like most of the stoves in those days; it could heat lots more water. The cold water came into the reservoirs

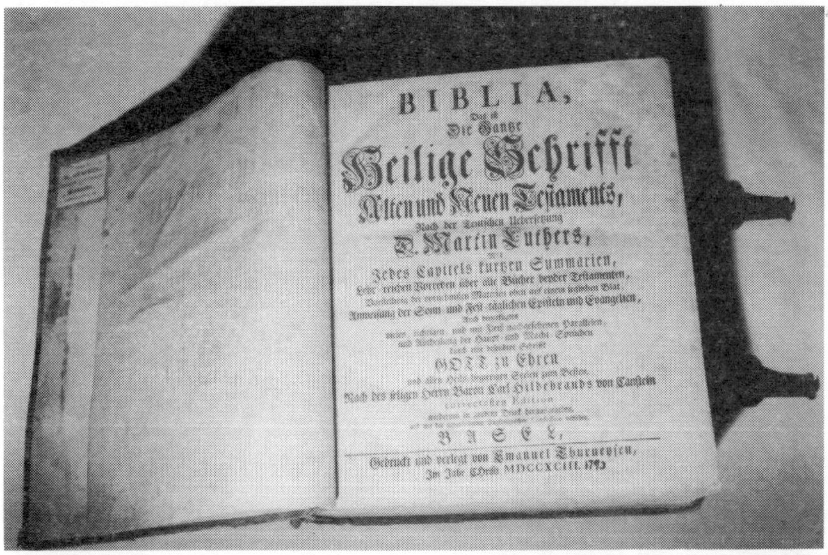

(TOP LEFT TO BOTTOM) **The German bible of 1793. The family's 100-year old secretary. Silver set with "wild rose" pattern.**

called water jackets and was heated by the firebox and heat from the stove. We had a hot water tank behind the stove, on the back side of it. We hooked it up to that so we could have hot water. It circulated without the use of a pump. If you turned on the hot water faucet the hot water came out of the top of the tank into the sink; just as it does today.

We'd hooked it up to an old hot water tank when we first got the stove. Finally, after a couple of years we could afford a new tank. Old Henry Lutz use to do plumbing. He came out and hooked up the new hot water tank for us. We didn't have any hot water after he hooked that up; we just couldn't get any hot water. We couldn't decide what it was. One day I decided, I'd just have to figure this out. I looked that over for a bit and found that Lutz had the hot and cold water coming out the bottom of the tank. The hot water was supposed to come out the top of the tank; so I changed that, then we had lots of hot water.

We had an indoor bathroom in our new house. We had a bath tub, a sink and a toilet, that was really nice to have, I'll tell you. Before that we had a two-holer and a wash tub.

We'd been working steadier than most anybody and we were getting ahead with Mama and Grandpa working too. These were "Prohibition Times," the government guys figured we had to be making moonshine to do that. Lots of people just made it for themselves. Some of them got so they could make pretty good stuff and they'd sell it. I never had anything to do with making moonshine, there were sure alot in the country doing that though.

Ed Johnson ran the store at Wren. Every time anything came up legal wise, a burglary or anything; they would get hold of him. They figured he would know where the guy lived. One time when I went in there, Ed says; "I don't know if you are making any moonshine or not. If you are, you better be mighty careful cause those government guys are watching you." Next evening, a couple of guy's showed up wanting to camp near the house. I wouldn't let them camp there cause Mama was there alone, that made them more suspicious than ever.

So Bill and I, we decided that we could have lots of fun over that. One night they came up to our place, we'd heard they were coming. So, I took off through the woods with a lantern. Oh, boy did they follow me. Those guys took off on foot right behind me; I ran clear over to Harris. That was several miles through there, probably five or six miles. They ran fast right through that thick brush; I took them through the worst stuff I could find. I ran in and out of creek beds with them following all the way. What they didn't know was that I'd arranged to have Bill meet me over there at Harris with the car. When they got over there, they didn't have a way to get back. Their car was still parked by the highway near our house when I got up to go to work in the morning. I should have let all the air out of their tires; I didn't think about doin' that until later.

I went up to Blodgett with Old Fred and Rex to look at some timber. Rex bought us some beer. That was the first time it had been legal for a long time. He got me a bottle. That was the worst tasting stuff there ever was. *Whew*! I couldn't drink none of that, couple of sips, that was it; so old Rex took it. Next time Ethel, his wife, saw me. She said; "George, what did you and Rex do last time you were out looking at timber?" When he came home, he got into the tub and stayed there, he just sang and sang; it was kinda like he'd had something to drink. When I asked him what he was going to do with the timber, he laughed and said; "It's so small, I'm just going to thrash it!" I'm sure Ethel figured it out, and gave him a bad time over that, Rex never did that again.

CHAPTER 9

WINTER LOGGING AT THE RAT'S NEST

Many serious logging accidents took place during the early 1900's. In the late 1920's through the early 1930's the logging crews George worked with, built roads along steep canyon slopes. Logging roads were built straight up and over the hills whenever possible. First, they began building the road-bed using picks and shovels, then plowing and leveling it lengthwise. The road-bed leveling was done with a Fresno. It was a big bucket, three times as wide as a wheelbarrow with a steel lip and additional steel on the back for weight. The bucket was two-feet high, eight-feet wide and six-feet long. All of this was resting on heavy steel runners. This equipment was originally pulled along by horses but after World War I there were more mechanical means available to do this work.

During the depression, Rex hired George to build roads for him. It only took two men to do this because George would pull the Fresno along with one of those early gas powered cats. The other man working with him, would hang onto the back of it to keep the lip up, out of the mud. George did his best to get that guy muddy. He was exceptionally good at using his surprised-innocent look at exactly the right moment. Few men could detect that this look was put-on for their benefit. They'd have killed George if he hadn't flashed this look of innocence just as they were being thoroughly muddied.

If the road had to run along the side of a hill they used a crowd to push the dirt from the higher to the lower part of the road. The blade ran at a forty-five degree angle to the back of it. Their job was to crowd the dirt toward the lower side. They made certain there was always a slight slant toward the outside of the road-bed. This kept the water from standing on the road. The steepest part of the road was then covered by a plank road.

After McCoy's mill in Kings Valley burned he'd built another one. This was over near Camp Adair at Soap Creek. There were a couple of miles where the Soap Creek road was real steep. McCoy paid his crew to build a plank road in there. They started by placing logs down into the earth and flattening the top-side only. There were two parallel tracks of these logs running end to end down the hillside. Covering the tracks of logs were planks three-inches thick by twelve-inches wide about ten-feet long. The men kept the planks from working loose by snugging each tightly to the previous one and attaching it with spikes. Additionally, every six-to-ten feet along the way, the men cut deep five-inch notches into the logs and placed cull rail-road ties into these, this helped keep the planks in place. These plank roads were often times, built along a twelve to twenty-five percent grade. The rain falling on these ran-off sideways due to the road bed's slant, keeping the road-bed from washing away.

They were logging the Link Allen and Henry Plunkett Canyons and the Ed Johnson place. George worked on many dangerous jobs in those steep canyons between Wren and Kings Valley. To make it safer for the drivers, the men selected a large tree stump, just above the steep-plank road and ran a cable around it. They ran an old one-inch sky-line cable the full length of the plank road. There was an eye on each end of the cable that could hold twenty to thirty thousand pounds. During those years they had one and a-half-inch hooks on the back of their trucks. They would stop their truck at the top. Here on level ground, they would hook onto the eye of the cable just before they began their decent down the steep plank road. The weight of more than one quarter mile of this cable was enough to control the loaded logging truck. This acted as a brake on their way down with a load. These stumps would soon have a deep groove in them. Moving the cable to a different place each time it wore deeply and eventually axing away that part allowed them to use the same stump for several months.

During these early years, the trucks had only three-inch bands for brakes, no loaded-logging truck could rely on those brakes. George devised an additional hand brake for the truck. He used a pulley from the rear end of a Model-A. This came up into the center of the cab where he'd cut a hole for the hand brake. George drove this by himself most of the time.

> It took four hands to drive one of these logging trucks with that hand brake down a plank road. Two-hands for this pulley-type brake; a third-hand to hang onto the stick-shift, since the transmission tended to jump out of gear; and the fourth-hand to steer the darn thing. The worst thing was this pulley-type hand brake would slip, loosening at every bump; you always had to watch it. I ran that for a several months without any problems.

George never went across an old bridge, or down a plank road, without previously deciding what he would do if the bridge were to break, or the truck were to get out of hand. Rex always said; "Anything dangerous was a sure thing for George."

Lyle Doughtery helped George drive that logging truck with the hand brake several times during bad weather. Rex wanted George to work down at the coast for him so George taught Lyle to drive that truck down the steep-plank road. Rex hired two men to take George's place. Lyle and another guy drove it, one man ran the brake and the other steered and kept the transmission from jumping out of gear. Lyle died later that same year from an accident that happened while they were unloading the logs from his truck. A large log rolled onto him killing him instantly.

> We sure hated that. Another man was hired to drive on that plank road. One time, he didn't stop soon enough to hook onto the skyline; he needed to stop aways from the plank road, up where the ground

was level. The truck's brakes were overloaded, it got away from him and went over the side smashing the truck real bad, lucky he wasn't killed.

McCoy was having trouble getting anyone who could drive that steep plank road he'd built over there at Soap Creek. The drivers were having lots of accidents that road sure was slick, real dangerous. Some got hurt pretty bad and tore the trucks all up. It was about a twenty percent grade, mighty steep all right, it wasn't a safe job.

Later that winter McCoy offered me six dollars a day if I'd drive for him on that. This was two-dollars a day more than Rex was paying me. So, I went to work for McCoy. There was only one truck bringing logs out of the woods. I drove up and down that steep-plank road all day long; made ten loads a day. I worked about ten hours a day driving for him that winter. Trucks in those days didn't have much of a brake, gearing down was the only way sometimes. This was dangerous because the bands would break, leaving the driver with no brakes. These were single axle trucks with rubber tires and they only had one tire on each side. Now they've got four tires on a side. I drove that for him during the rest of the winter; never had an accident. I hauled there without any trouble and kept his mill full of logs. Never let him run out that I knew of. I worked for McCoy for about six months that time.

I'd promised Rex I'd drive for him when his new Dodge Graham truck and trailer he'd ordered came in. Rex offered me fifty cents an hour to drive that for him. He'd bought that new truck, they called it a big-four. The engine had four cylinders, each cylinder was twice as large as most cylinders, so it had twice the power of most four-cylinder trucks. It was in May of 1929 when I drove that for him. That month I got a check for two-hundred dollars, for driving for four-hundred hours, that's a lot of hours! Even in the winter I often worked into the night hauling logs, sometimes I'd even load by moon-light. We didn't haul logs on Sundays, only once in a while. Sometimes I'd work on the truck on Sundays. Boy! Mama and I thought we were getting rich with that two-hundred dollars. Mama saved most of that, it was alot of money, I'll tell you! I thought maybe Rex would kick on it, he didn't. He said that was some of the cheapest logs he was getting. Two-hundred dollars was an awful lot in those days.

One night, Mama and I saw McCoy in Corvallis on the street as we were going to the movie. McCoy asked me to come over and see him the next day. He said; "I'd like to talk to you about some business I'm planning on doing." Next afternoon, I went to his house to talk to him. There were lots of people there. When I went in, I saw Mrs. McCoy was crying, he'd died just after getting back from church that

morning. Later, Mrs. McCoy told me he was planning on offering me a partnership running the woods for the sawmill he was planning to build in Roseburg. Mama and I probably would have moved to Southern Oregon if he hadn't died.

During the winter months Rex never expected much of a profit. His men couldn't work at the coast where he made more money, logging there was impossible due to poor road conditions. The depression was setting in and families were struggling. He began realizing he needed to provide year around work for more of his crew that winter. The hard working, conscientious men were concerned for the well being of their families.

Rex decided to build a new mill in the area known as the "Rat's Nest." It was an old mill site overrun with rats. It was six-miles south of Kings Valley and four-miles north of Wren. This was about five miles from our place in Gellatly Canyon. Rex kinda put me in charge of building the new mill. I'd helped Grandpa Hunter build McCoy's Mill, so I knew some about building a mill.

When I went in that first morning, Rex and I looked the place over and talked about what we thought might be best. I'd helped Grandpa Hunter build the mill at Kings Valley for a couple of years while I was still in high school. I had a good idea about how we could make it easier on the men working at the mill to load the trucks. After Rex left that afternoon, I kinda started moving things around getting ready; all the other men were mostly twice my age. They just kind of ignored me, didn't ask if they could help. Rex came back later that evening. Most of the guys had gone home. I told Rex I didn't think those guys were gonna work for me. Rex said; "They will in the morning." He sure was right, they worked real hard after that.

When we were working on McCoy's mill, we used logs for mud sills, the foundation of the building. We put them in length ways then built our posts on top of those. We made those posts about eight-feet high so you'd have plenty of room to walk underneath; some mills had their pinions down low. The reason we raised the mill up was to have a place to store the lumber. By storing it at a level lower than the level the men were working at; it kept the men from having to lift every piece of lumber up above their heads. The sawmill at the Rat's Nest was finished enough to run in two months. It cut fifteen to twenty-thousand board-feet a day. We were logging second growth, these were fifteen to forty-inch diameter logs. Rex Clemens worked his men there in the winters after that for about ten years.

After the mill was built, Rex hired me to run the cut-off saw. Well, that was a pretty dusty job. I never liked it. I cut the lumber to the right length when it came out of the mill. If they didn't cut the logs right, I'd have to trim it back. Most of them did allow some for trimming. When you're sawing big logs, it changes the cut and throws sawdust all around. I hated that, I itched all over; that just drove me wild! Most of the saws were overhead saws, they had a weight, an old rock or pulley, to hold them back. The man running that saw could lose his fingers if the cable broke. Quite alot of people got their fingers cut off, that was dangerous work in those days.

Rex came in one day and he said; "I'm having trouble getting the logs out of the woods so you make up your mind whether you want to stay in the sawmill or work in the woods." I said; "I'll take the woods!" I've been logging since that day.

Rex had been using horses up to that time, Bud Reynolds and Orville Edwards logged with the horses. A couple of other men ran the gas powered mechanical loading donkey to load the logs. Others did the falling and bucking. He decided to get a little cat, figured I could run it. He bought a Cletrac 30. It cost about three or four-thousand dollars. I started using that in the woods to build roads and clear root wads, it worked pretty good. I thought it needed a blade though, so I got a big piece of iron and attached it to the front of the cat. It didn't move at all, you couldn't lift it up or anything, but it worked real good.

This chicken farmer lived close to the mill. He could never understand where we kept the cats and donkeys he heard us talking about. One day, he said, "I've never seen those cats and donkeys that you guy's are always talking about. I've never once seen you hauling 'em anything to eat." He'd moved up from a city in California, we got a big laugh over that.

That acre pond there at the Rat's Nest would fill up with silt in a year's time from the logging done above it. Each hard rain would bring in lots of wood chips and bark along with the muddy water from above. The pond would fill with so much mud the logs wouldn't float. It had to be cleaned out when that happened. Sometimes it would dry up and that would make it hard to get the mud out in the fall.

I took the little Cletrac 30, with my homemade dozer blade on it. I took the fan belt off it and turned the intake pipe up so no water could get into the crank case. I drove that Cletrac 30 right out into the middle of the pond to scoop the mud out. I cut a hole in the dam, right there in the middle. I'd take a swipe with that blade on the cat. I'd push it right toward the opening in the dam. The water was coming up all over the floor boards where my feet were. The flow of the water running through there would wash the mud right out. I was slushing that right out of there. Rex, he comes up there and he says; "You've ruined that cat. But go ahead, you've already ruined it." So, I just kept kicking that mud out. Old Rex thought sure I'd ruined that cat.

We couldn't have gotten it out any other way. We'd have had to work for days with a donkey and a drag line. The crew at the mill wouldn't have been able to go to work for quite some time. I could move more mud in an hour with that cat than we could move with a donkey and a drag line in a week. That night I plugged the hole in the dam; the next morning it was full of clear water. Rex could put his men back to work. The only thing I hurt was the brushes on the starter and generator. If I'd sealed that up it would have been all right, I didn't think of that. The cat ran till the new brushes for the starter and generator came in. It didn't take long to replace those. After that, it was good as new!

You'd had to have water in the pond to float the logs up to the mill, or you couldn't have run it. That pond was one of my biggest troubles. Even when I worked out in the woods, they expected me to take care of that pond. The pond had eels in it and they would tunnel under the dam. I don't know where those eels came from. It was my job to fix the dam. Lots of Sundays, Bill and I spent our time plugging up the holes in that dam. I would get Bill to help me, he was good help and always needed the money.

The pond had a dirt dam with stakes in it; sometimes the pond wouldn't have a drop of water in it. We'd find some timbers and some gunny sacks. We'd mix dirt, straw and manure together and put it into those gunny sacks. We'd use those to plug up the holes, if you just put dirt in, that loose dirt would wash right out. You may need to put six or seven of those in that hole and use those timbers to help hold them in place. If the hole wasn't too big, you could put alot of dirt in on top of those and you could seal it real good. It would take six to eight hours for the pond to fill up.

Rex told me; "I may not know how to get something done, but I know how to get somebody who can do the work." Rex would just say; "There she is, I gotta have logs Monday morning." He didn't care how we fixed it; he just told me; "Fix it!" That's all he ever did, he never knew how, that's what he hired me for, because I could get it done. Whatever Bill and I had to do to get that water in there, we did it.

While we were at the Rat's Nest, Rex hired Buck Arbuckle, he was the youngest worker there. He was only sixteen years old then and wanted to be a logger. When he first came to work he didn't have much muscle. The guys there would take turns wrestling with him just to

toughen him up a bit. By the end of that year he'd learned to take care of himself and his muscles were stronger. We liked him real well; he worked hard and joined in on the pranks.

One of the first pranks Buck and I played was on Old Chink Simmons. When Chink worked in the woods, he usually bucked logs. Lots of times this old guy sharpened Rex's saws. He lived up by Ridenour's Station there in Philomath. This morning he was out in the woods with the crew. He was working up ahead of some of us guys. It was early May and we were logging up in that country, up there behind the Rat's Nest.

It was kind of a drizzly morning; the water was warmer than the air outside. Well, Buck and I were working together and we noticed this little garter snake. Buck caught him up and I walked over and got Chink's water jug. Buck put the snake into that water jug and I put it back where I got it. That water jug frosted up and you couldn't see the snake from the outside. This guy was always drinking from his water jug. He picked that thing up and took a drink, you know. Pretty soon that old snake stuck his head out, he had his tongue out. Old Chink never jumped, never moved at all. He said; "I thought I had that plugged up so nothin' could get in there."

Old Chink never even flinched, drank that water with the snake right there in it. Buck thought that was pretty good. The best part of getting a joke played on you was not to let on to anyone that anything unusual had happened. Buck started helping Chink get that snake out of his water, that darned old snake didn't want to come out. Buck placed the water jug near the fire. After a while, the water got warm enough and that old snake came out on his own. Buck's mother always said that she didn't mind making his lunch but she sure hated to open up the box at night to wash it, there was apt to be something in there, a mouse or a snake.

Buck thought blasting was pretty exciting. Every time I went to blast he wanted to watch. We were out at the Rat's Nest and I was logging up on the top with the cat. We had a swing tree and we swung the logs down about a thousand feet. Vip Gee ran that old gas donkey we were using for loading those logs. When the gas barrels were empty we would send them down on the rigging and they would get beat up. Buck was working down below as a chaser. One morning he said; "Rex is taking on about those barrels being all beat-up because they charge him extra for those. Do you think a stick of powder would straighten them out?" I told Buck I thought it would, never thought about him using it. I'll be darned, if he didn't take the plug out of one of those, and drop a fuse and stick of powder down there. It straightened it out all right, just like a piece of sheet metal. By gosh it

sounded just like a bomb going-off. Rex came running out of the office. He said; "Who told you to do that?" Buck said; "Well, George said it would straighten it out." Good Lord! It sure did!

We did our first yo-yo logging up there in Link Allen Canyon. We used cats to let the trucks, loaded with logs, down a steep plank road. Sometimes we'd do a little logging in there in the winter. It was awfully muddy there in Link Allen Canyon. One time we were having a mud ball fight to keep up the fun. Buck put a rock into his mud ball and it broke out a truck window, he didn't do that again. Sometimes, Buck would drive out there at night to drain one of the cats. He'd let his little brother Henry ride along with him pretty often. Buck was very mechanical. One night he took a malfunctioning voltage regulator home. He said that his little brother stayed up late, eagerly watching him as he took it completely apart and put it back together again. Next night after work, Henry was waiting for him when he came home, he wanted to know if the regulator worked when Buck put it back on the cat. It did, he was pretty good that way.

First tree I ever climbed was the one that was standing right outside the office at the Rat's Nest. That tree was used to move logs from the trucks into the pond. They left that standing there, it wasn't straight enough to saw. Ethel, Rex's wife, worked in that office, she did all the books for his logging business. Vip Gee, her uncle, and Cliff Gee, her oldest brother, worked in the woods. Fred Gee, Ethel's father, usually did the high climbing but he wasn't around.

There was a block up there in that tree and we needed it moved so I decided I might as well get up there and move it. So, I put on the belt and spurs; thought I was ready to go on up that tree. Cliff was running the pass line, taking me up there. The tree was over one-hundred feet high; it was crooked and had bumps on it. I didn't have any experience climbing, didn't know how to kick out from the tree to free the line from rubbing my hands against it. That tore the hide right-off the backs of both my hands, even my fingers. It hurt awful bad. I remembered some words I'd heard, and started using some of them. I came down some better than when I was going up. Ethel was in her office there at the mill when I went in to get something to put on my hands. When I came in, Ethel said; "George, I didn't know you knew those words!" I guess she thought I was learning to be a preacher.

"Gees"
Vip Gee: Rex's wife Ethel's uncle
Cliff Gee: Ethel's oldest brother
Fred Gee: Ethel's father

CHAPTER 10.

THE DEPRESSION YEARS

The Depression lasted ten years and caused unbelievable pain and suffering among those not rooted in the land. If you had a little land you could grow your own food. Without that option, you were destined to experience poverty as never known before in the United States. Thanks to the wisdom of President Roosevelt, he developed work projects that gave people food to eat and work to do during this period of extremes. George recalls these times and the effects they had on himself and the people he knew.

Just before the depression set in, Gail was working for the State of Oregon, running a highway maintenance crew. He was getting four-dollars-a-day, I was getting two-dollars-a-day working for Rex. Can you imagine that, that was a day, not an hour. Just about everyone's wages were cut-in-half when the Stock Market crashed in October 1929. The worst part of the depression was during 1931 through 1933. Back then, everybody wanted to work, if you didn't work you didn't eat.

I was working on the rollway hauling logs. Orville and I got a dollar-and-a-half-a-day; all the rest of Rex's crew got a dollar-a-day when the mill was running. Those old mills were kinda thrown together; they broke down alot. If it broke down, we fixed it; usually we didn't get paid for that. You did that on your own so you could get paid the next day or the day after that. I remember a good month's pay during the summer was twenty-eight dollars. Winter time was a lot worse, Rex ran the mill at the Rat's Nest on a break-even budget then. He did this so his men would be able to feed their families. We wouldn't work every day, cause the old mills broke down so often.

If the mill wasn't running, there wasn't any money to pay the men with. Pret' near all the small sawmills went broke at sometime during the depression. During the worst part of the depression Rex was getting seven-dollars for a-thousand-board-feet, most of the time it was ten-or-twelve-dollars a-thousand-board-feet. That included falling and bucking, trucking to the mill, milling and trucking on over to the railroad yard.

There wasn't much you could do back then. You paid your rent, bought kerosene for your lanterns and gas for your car. You had to buy some staples—flour, sugar, salt and stuff like that. We paid rent to Old Walt Kisor, started out as five dollars a month, during the worst of the depression he cut it down to a-half of that in 1933. That was about twelve percent of your check, you know. Mr. Kisor came out and suggested we cut it, he knew people were having a hard time. Everybody was, and lots of people weren't paying him at all. Mama said; "He was kinda of an old grump back then, but he had a big heart." He was helping lots of people; especially young married people. They would go in to his store when they got married and get a stove, a bed and stuff from him. He'd usually give it to them on credit, they would pay for it later.

Mama's garden and livestock gave us pret' near everything we needed for food. We didn't make much money, but it didn't take much then. You could buy a loaf of bread for ten cents; now it's a dollar-and-a-half or more. We had to buy some work clothes and shoes. One time when I bought a pair of cork shoes they cost eighteen dollars. That was two-thirds of my pay check that month, that was a pretty big expense. If you didn't have those cork shoes you might fall off the logs. We were working eight-to-ten-feet off the ground, that was pretty dangerous. If you were working all day long on the rollways, a pair of those would last you about a year. Later, when I started hauling logs and stuff, they would last me two years.

Well, those were pretty difficult times back there in the Thirties. We never went hungry, I always had a job, some people didn't have jobs. I always worked lots of hours when I worked for Rex. He never paid overtime though, I didn't get time and a-half, just got more hours. There was no unemployment or health benefits in those days, they didn't have such a thing as health insurance.

George and Helen were living out in Gellatly Canyon which was on the highway to the coast. There were numerous tramps during those years. Many of these unemployed homeless men were starving, so people fed them whenever they asked for something to eat.

During the depression, tramps would come by, sometimes several every day. They would stop and get something to eat from Mama. She was concerned for her families safety so she would send them on their way soon as they finished eating. We never had any trouble with them. One Sunday morning, two men came by dressed in suits, I didn't have any nice clothes at all at that time. Those guys had real nice suits on, so I wouldn't give them nothin' to eat, I made them go on. I figured if they could afford clothes like that, they didn't need anything to eat. Mama thought I was being kinda tough on them, as far as I know that's the only ones we didn't help.

There weren't many dangerous people around in those days. We'd see some drunk people along the roadside once in a while. Some tramps, but not people around who would harm you, shoot you. There weren't drugs then like there are now. Sometimes we had hitch-hikers come through. Back then, if anybody wanted a ride; you'd pick them up.

You shouldn't do it now, though. I liked to pick them up and give them a ride, never had one pull a knife or gun on me.

There was one guy that I thought had a gun, he kept his hand in his pocket. When I got to where we lived, I didn't stop there, I kept driving. I drove all the way to Blodgett, about four or five miles past our place. I drove right up to the store, it was all lit up with lights. I told him that was as far as I went, he got out and I turned around and came back home. I was afraid to let him out at our place. He could have broken into the house or the service station during the night or he might have waited until the next day, after I'd gone to work. He was a different kinda guy, I didn't trust him. It wasn't long til we heard a guy was in a little house up at Dickinson's. One day when the cops went by; he took-off running, they caught him and took him in with them. Come to find out, he'd killed somebody down in California and broken out of jail before he came up here. I always wondered if he was the guy I picked up on the road and hauled over there to Blodgett.

Even during the worst part of the depression we'd have several people stop each day at our service station. The cheapest gas we sold was about seven-or-eight cents a gallon, when there was a gas war. As a rule, it would be sixteen-or-seventeen cents. We'd sell tires once in a while, tubes, gasoline and oil. Some of the people would charge up a big bill. Next thing you know; they'd move out of the country and never pay us. We sold cigarettes, candy and ice cream, never sold any beer. When it became legal, some of the neighbors wanted us to put that in. I never wanted to be responsible for somebody getting drunk and driving off the road and killing somebody. We didn't want people drinking and driving so we wouldn't sell it. We figured that would stop people from drinking. We could tell a big difference in our business once beer came in. People would stop at other places, because they could get beer. We didn't sell as much gas after that.

Often, society discredits ideas of significant importance. George and Helen made the leap of responsibility, that today -- society is only beginning to comprehend; as reflected by the legal system. The sale of some products holds with it the responsibility of the consequences of use. Individuals still drink and drive, now the courts hold both the drunk driver and business responsible. Many people have lost loved ones through the criminal negligence of selling alcohol. People in general should not try to please their neighbors. They should instead, aspire to be conscientious about the effects of their actions. George and Helen gave up their possible profits to assure the safety and life of their loved ones and others. A much greater treasure!

Once in a while someone coming by would have car trouble. If they broke down at night, they wouldn't have a way to get home. So, I'd take them home if they needed me to. One of those times it was a couple from Harlan, a girl sixteen years old married to a man three times her age. I thought she was too young to be married to this old guy. Their car broke down by the service station. They'd come up to our place to ask for help. It was already dark out, it was winter time and there was lots of snow on the ground. I took them home cause we had no place much to keep them. I remember that the girl said she'd had never seen the lights of a town before, she thought it was a great place. In those days there were many kids that were sixteen before they ever got into town. She'd bought a package of gum, she opened it, gave me a stick, took a stick and gave him a stick. She was going to save two sticks for the Saturday night dance.

Mama usually ran the service station, this one evening I was home and Mama was busy feeding Lucy. So, I ran down there to see what this guy wanted. There was a sixteen-or-seventeen year old kid wanting some cigarettes, I got him some and went to get his change. There wasn't any money, the change was gone, someone had gotten it. I figured he had it, that he'd stolen it. I grabbed him and turned him upside down, the money came out of his pockets all over the floor. I turned him loose and he went running home.

Pretty quick his sister's boyfriend came down to beat me up for turning that kid upside down. This boyfriend was a boxer in the Marines, he was about as big as I was, but he wasn't as heavy. He got into a fight with me. I remember I'd been working on the pond up at the Rat's Nest and I had on hip boots. Those are not very good things to be wearing in a fight. Mama heard the racket and came running down. I could hear her yelling; "Kill him George. Kill him!" He bit me on my thumb. It surprised me and I hit him hard; knocked several teeth out of him, beat him up pretty bad. Do you see that? That's where I got that big joint here. I've never had many fights, most people didn't want to fight me. I was in good shape then, and I always weighed about twenty pounds more than I looked, that gave me quite an advantage.

We worked at the coast during the summers, sometimes I'd be away for two or three days at a time. I'd drive back from the coast on Wednesday or Thursday night and again on the weekend. It was too far for me to drive home every night. Bill and I were worried about our families, they were alone so much of the time. We were afraid of someone coming around when we weren't home. Often, Mama, the kids and Grandpa, were home alone for days. There were lots of old mill shacks, whenever people moved out of one, the neighbors would burn it to the ground. They did this to keep poor people from moving in there again. They thought those poor people might steal things if

George's 27-inch block plane.

they didn't have any money. One Sunday, Bill and I burnt down a couple of old mill shacks. We were trying to protect our families. In a way, it seems like burning down those shacks wasn't the "right thing to do."

We had lots of snow in the Thirties there in Gellatly Canyon. I would always help our closest neighbors the Muller's shovel all that snow off their barns and sheds, they lived about a mile away. We had one real big snow storm, that was in 1937, that was the biggest snow storm we ever had. It started early one Sunday morning, we were all out there working, trying to keep up with that snow. It snowed steady all that day and all that night. It just kept coming right up, we had about thirty-two inches before it quit. That storm broke the Muller's barn down; it took lots of buildings down all over the country.

Gail was foreman of the highway crew that was responsible for the road in front of our place and on up to Eddyville. He had his crew work right straight through all that first day and night, he had his crew grading the road constantly. He was the only guy in the country who had his section of the road open. That didn't do much good though, cause no one else had kept their part of the road open. You could only go from our place in Gellatly Canyon to Eddyville, couldn't get through to the coast or to Philomath cause the others weren't able to keep their part of the highway open.

(LEFT) (TOP ROW) Helen, Mabel and Oliver; (FRONT) Joyce, Lucy . (ABOVE) Lucy and Joyce at their new house.

We'd bought a new Chevrolet that year. We kept it in the garage and wouldn't take it out because of that snow. That was our first new car. It was a 1937 Chevrolet Coach from Clark and Haight's in Philomath. We bought it so Mama could get the girls to school. It would be Joyce's first year of school. We wanted a car that was dependable and safe for the kids. It had front and back seats and only two doors, we wanted this so the kids couldn't fall out. The car we wanted cost eight-hundred dollars. Mama and I had saved six-hundred dollars and Rex offered to loan us two-hundred dollars so we could pay cash for it. A heater was going to cost thirty dollars more. We thought that was pretty high so we ordered it without a heater. When the car arrived at

Clark and Haight's garage, they put a heater in. We told them they would have to take it out because we couldn't afford one. Since that was their mistake, they said they would put the labor in for free. They talked us into buying the heater at cost, that was fifteen dollars. Our 1937 Chevrolet served as the family car until 1946. After that, we used it as a crummy for my logging operations.

Those little girls were warm and cozy on their way to school that winter. George and Helen were glad they had the heater in the car. Since they were very careful with their money, they used batteries as long as they possibly could. After a couple of years, occasionally the car wouldn't start so Joyce and Lucy would have to run the two-miles to school. George being so ingenious, figured out how to charge everyone's batteries.

George and Grandpa had built a water wheel to provide electricity for their home, shop and service station. There was a water wheel building beside the creek that housed the large generator. This was a small ten by ten building with a nice wooden floor and a couple of windows, it even had a good roof. He added a second belt to the water wheel's axle, to this he attached a six-volt generator he'd taken from a car. Now he was able to recharge his own as well as his friends batteries.

Gail's crew turned their trucks around at our shop. If I was home working on something when the weather was bad, I'd have a fire going and they'd stop in there at noon to eat their lunch. I'd build the fire up real good so they could warm themselves.

One time when my brother's crew was stopping in for lunch pretty often, there was a guy that would go off by himself at dinner time. He was a new man that Gail had just hired from over near Eddyville. He wouldn't eat with the rest of us, so the guy's got to thinking about that. This one time, Gail sent him someplace else. Those guys looked in his dinner bucket, he had nothin' in there but potato peelings. They'd had boiled potatoes for the family the night before. He'd brought the peelings for his dinner, that's all he had, that was pretty bad. He didn't want to let anyone know about that, he had to be a pretty good guy. We got a bunch of food together and took it out to him and his family. I remember Mama gave them lots of jars of canned meat, fruits and vegetables. Mama was real generous, she knew people were having hard times. There were lots of families suffering in those days. There wasn't one of us had any money to speak of.

Mama and I used to play cards with several couples. We'd take turns having people come over. Sometimes they'd play at our house; at other times, we'd play at one of their houses. One night we were playing cards at Bill and Susie Muller's house. I went in the kitchen

for a glass of water and I noticed the alarm was set for four o'clock in the morning. They had quite a few cows that Bill and the kids milked early, before breakfast. So heck! I just turned the alarm hand back to two o'clock in the morning. Next day, I learned that they had a hard time that morning, the cows didn't want to come in and they didn't give much milk, at that hour. A couple days later, when I saw Susie, she said; 'I'm gonna come over and get you up at two o'clock in the morning!' I said; 'Susie, you should've known as much as the cows.' She said that they didn't figure it out 'til the news didn't come on the radio when they were eating breakfast.

After the girls were born, we didn't want to take them out at night very much. So, friends would come over to our house. We'd always have dinner, then they would stay and play cards. Mama would make up lots of candy. We'd eat that while we were playing. After we finished playing cards, we'd usually have pie or cake and some ice cream. The ice cream was a real treat. We had a little ice box in the service station where we could keep it. Most people didn't have one, in those days. So, most of them didn't have ice cream.

Yes, we had lots of fun when we played cards. When we played on Sunday nights we'd play cards til about one o'clock. Only thing, when everyone was at our house, some of them wouldn't go home at night and I had to get up early to go to work. Sometimes, I'd say; " Mama let's go to bed so these folks can go home!"

Frank and Alice Nordyke, our other neighbors, moved up over the hill between Bill's place and ours. Mama and the girls use to walk over the hill to visit them when the girls were pretty little. Then, when Joyce was about five years old, Grandpa finished our new house, we moved right in there. It was by the service station, that made it easier for me to watch it. Nordykes moved into Kisor's old mill shack where we'd been living. While they lived there close, we played lots of cards in the evenings. Mama would make up lots of candy when they were coming over. We played cards on Saturday and Sunday nights.

After a few years, they moved to Burnt Woods, I remember going up there to visit them once in awhile. Frank was always telling me about how old and lazy their horses and pigs were. I'd told Sonny, the oldest boy, that I had some high life and it would give those lazy animals some spirit. He wanted me to bring some up to his place so we could put it on their horses. So this one Sunday, while the rest of them were still in the house, Sonny and I went out in the pasture after dinner to see their horses. They were all lying down, they were real good pets. They never even got up when we came out. We put some of that high life on those horses' backs. It made them go kinda crazy. They jumped up and they stretched, they kicked up their heels and away they ran, just like a bunch of yearling colts. Old Frank, he came running out and he wants to know what-in-the world's wrong with those horses. I said; "I thought those were old horses, those horses are running around like colts." He says; "I just don't know what's gotten into them." Frank went back in the house, so Sonny and I went back into the barn. Underneath the barn Frank had several sows with litters of pigs. There were some big cracks in the barn floor so, we got above some of those sows and poured that stuff down on them. Good Lord! They came out of there just a squealing and running around like crazy, that Nordyke kid had more fun that day. Probably more than he'd ever had before. I don't think Frank ever found out. I never told him, I'm sure Sonny didn't tell him either.

Vip, Cliff and Fred, Gee (left to right) moving the loading donkey across the Siletz River.

CHAPTER 11.

LOGGING SUMMERS AT THE COAST

Rex found work for his men in 1931 and 1932 logging heavily timbered areas of Sitka Spruce along the coast. His crew worked there only during the summers. Logging was first done by this crew at South Beach, the area across the bay from Newport. This was the men's first venture out of the Willamette Valley although certainly not their last.

Before 1914, logging crews considering Sitka Spruce as undesirable, began avoiding it whenever possible. A surprising development took place during the First World War. Sitka Spruce being light weight and strong became valuable for airplane construction. Recognition of its value had significant implications for the logging industry in Oregon as there were heavy stands along the coast. After the war, they began using it for fish boxes. Finally, men realizing this wood's value began using it for boxes for storing delicately flavored foods.

There was a butter tub manufacturing company a few miles up-river in Toledo. These tubs were very large weighing five-hundred pounds when filled with butter. Spruce did not have any pitch in it, therefore the butter packed for export to Europe and other countries would maintain it's sweet taste. There were many dairies along the coast supplying the cream needed to make butter.

George and Fred moved the cat and the loading donkey to South Beach. Early one spring morning they loaded the equipment onto trucks and drove down to the bay at Newport. That was before the building of the Yaquina Bay Bridge, so they loaded the equipment onto a barge. This was quite an undertaking. They were concerned that the barge floor would not be strong enough to hold the weight of the loading donkey. So, they brought their own large timbers for the transfer. These timbers were eight-inches by twenty-four-inches by thirty-feet. They used these timbers to transfer the six-thousand-pound loading donkey and it's sled to the barge. Then they moved the Cletrac 40 tractor onto the barge where it was securely chained. The barge had many large rings to allow securing of the cargo for the trip across the bay.

> **After unloading the equipment we drove south along the coast about two miles, then we turned left on a plank road and drove up to where we would be logging. Rex hired me to run the cat as the trees at the coast were too big for horses to move. So, we logged at the coast with this new Cletrac 40 Rex'd bought. My job was to build roads and log. Fred was the woods boss, and he was real good to work with. He always talked it over with us before opening up a new area. We'd figure the best way to log an area and where to build the road.**
>
> **While we were logging up there, we stayed in cabins on the beach side of the road, only about five-hundred feet from the ocean. I liked to**

listen to the sound of the ocean at night, it was relaxing after I'd worked hard all day.

They worked a couple of years at South Beach. Rex had several cabins for the crew to stay in while they were working there. Rex and Ethel stayed at the coast occasionally. Fred, his wife, Lillian, and Hazel their youngest daughter stayed in one of the cabins. Hazel was only fourteen years old at this time; she later married Stanley Lowther. Cliff, his wife, Irene, and their son had a cabin also. Most of the time Ethel's Uncle Vip and his wife, Pearl, and their daughter stayed in Philomath.

George would drive to South Beach early Monday morning, leaving home between four o'clock and five o'clock, depending on the weather. Driving to the coast in those days took longer, roads were poor and vehicles were slower. Wednesday evening he would drive home and Thursday morning he would again rise early for the drive to the coast, working three more days and returning home Saturday afternoon.

George lived with Fred's family when he stayed overnight at the coast, he slept on a cot on the back porch. One night, Fred and Lillian kept hearing George chuckling to himself, out there on the back porch. Finally, Fred got up and went out there. He wanted to know what George found so entertaining. George told him about a prank he and Cliff had played earlier that evening.

When Cliff and I were coming back on the ferry that night, there was a couple of guys, they were all slicked up. They had some beer and melons in the back of their pick-up and they were talking about this date they had tonight. We thought we'd have a little fun, figured we'd fix these guy's. I found a large four-by-four-by-ten-foot timber and I put that in front of their front tires. Cliff saw what I was up to so he kept those guys busy talking, the ferry pulled into the dock and Cliff and I got off. It took those guy's so long to figure out what had happened that the ferry started back with them still on it and that was the last ferry over that night. They never got to go on that date!

Sometimes our equipment broke down. We'd be waiting for the shop to fix it or for Rex to bring us parts. I liked the water and would walk along the bay. One day I was walking down there where some boats fastened down, one was a flat bottomed boat and it looked pretty safe. Thought I'd take a little ride out in the bay. I was out quite a way, when I noticed this fellow watching me and I figured it was his boat. I didn't know whether to go clear across the bay or come back in. So, I came back in and thanked him for the use of his boat. This Clifford Bartoe fellow and I got to be good friends. When I'd have part of a day-off for breakdowns, I'd help him crab.

Rex's crew was logging the Siletz area for the first time, summer of 1933. He was buying timber from the Siletz Indians and selling logs to C. D. Johnson's

Mobile bunk house at the Siletz camp. (left to right) Orville Edwards, Vip Gee, Perry Spencer, Fred Gee, Melvin Hawkins, Bill Adams (standing in front), Cal Tatom (squatted in front with Willard Green (in doorway). Three remaining men unidentified. The vehicle is a 1929-1930 Chevy.

mill in Toledo; C. D. Johnson's men were rafting them up in the Yaquina River. Rex had several bunk houses built; they called them "mobile mill shacks." Their dimensions were eight-feet wide, twenty-four-feet long, and about eight-feet high. The crew would set these up at the logging site each summer, they also set up some old ten-by-twelve-foot-canvas tents to hold more bunks. Each canvas tent had a stove, a wash basin, a mirror and four beds. When entering, the beds were placed along either side, with the head of a bed in each corner. George, Buck, Vip and Fred bunked together while they were setting up that summer.

> **Pack Rat's were always around our camps, running off with our soap and storing food in our boots. Old Fred was scared to death of those pack rats, he sure hated sleeping in those tents. One night Buck and I decided to play a trick on Old Fred, we figured we'd imitate a pack rat. So, we fastened a safety pin at top of the tent just over the middle of Fred's bed. Then, we ran a string through the safety pin with my cork shoe attached to the other end. Buck had a string that went from his bed to the end of Fred's blanket.**
>
> **After dark everybody went to bed. When Old Fred started snoring we were ready for him, Buck would jerk the end of Fred's blanket and punch the bottom of my bed with his foot; this would signal me to drop that boot right on Fred's stomach. He would toss and turn and hit at that old pack rat. About the third time, Old Fred came up a**

Logging at the Oregon Coast. Rex Clemens and a 1-log load.
The truck was a Federal.

cussin' he was gonna shoot that "Son-of-a-gun." When he saw those strings going to our beds; he knew what had been going on. All that noise woke Vip up, it sure tickled him, he really got to laughing.

One hot night I came in a little late for supper. There was this pony staked in the tent beside my bed. It had it's legs through two pairs of my pants, they even had one of my shirts around it. Someone was trying to get the best of me. I took my pants off that pony and I decided I'd have some fun, thought I'd ride that pony some. I wasn't figuring on going in but that pony just jumped up on the porch. So I thought, we might as well go on in and I headed him right through that cook house door. Orville yelled; "That guy will run right over us!" Boy you should have seen those guys pour out of the windows and doors. I made a circle around the table and there wasn't a soul left in there. I tell you I had the cook house all to myself except the kitchen.

The loggers stayed in one bunkhouse and the truckers stayed in another one. Bob Bennett was head trucker, then there were Hugh Pleasant, Slim Nutter and Lloyd Lutz. One night Old Lloyd Lutz came over there to our place. He was kind of an easy going guy, we all picked on him. He was quite a talker; he got to talking about how truckers were better than loggers. So, we took his clothes-off him, kicked him out the door, and sent him home. Old Bob came boiling into our camp and he said; "Nobody can treat one of my men like that!" So, we grabbed him, took his clothes off, kicked him out the door, and sent him back the same way. That shows you what the old loggers did back then. They were both good guy's, people needed to have fun, it was hard work back then.

The loggers and the truckers were always trying to get the best of each other. Slim, was always stealing pies or cakes out of the logger's dinner buckets, so we'd hide them. One day when I hid my dinner, I knew he was watching me; so, I went out behind some logs. I told Old Orville about it and he said; "We want to get even with this guy." So, he took my pie, smashed down the center and filled it with Tabasco sauce, then filled it back in with custard so Old Slim wouldn't see it. Pretty soon he went out, got my dinner bucket and came back holding that pie. I moved in close so I could watch him. He yelled; "Look what I got." He held the darn thing up and everybody looked at him. He took about half that piece of pie in his mouth, swallowed it and started to take another bite, then spit that out. Boy! He thought he'd been poisoned. He ran for the water jug, we always had a jug of water around. The guys knew what was going to happen so they'd poured it out, he couldn't get any water. He headed toward town, he was burning-up. On the way, he stopped where Al Daughtery was working on the road, Old Al wouldn't give him any water either. Slim had to go all the way into town before he could get any water. That stopped him, he didn't steal any pie after that.

Rex sold logs to C. D. Johnson down there in Toledo for several years. One year his union men went on strike. The men in our crew weren't in the union, we needed to feed our families, we wanted to keep on working. Those guys would sabotage equipment, and throw rocks at the gyppo loggers trying to unload logs at their mill in Toledo. They wanted to force it to close.

Rex said; "Well, we are going to have to go through that picket line; we'll take all the trucks in at one time with Bob in front followed by Hugh, Slim and Lloyd." Now, Hugh was from the South, he'd grown up with fights being common. He was a real stout guy, pretty stern and determined looking. Rex gave him a ball-peen hammer and told him not to be afraid to use it. Hugh laid that hammer right there on his dash, he'd lift that up at times for emphasis. They rolled their windows down to keep them from being broken. They were able to deliver the logs to the mill to keep it from shutting down.

One morning our crew found that sabotagers had crept in during the night. They brought in axes and cut the line holding our boom, this let all our logs go down the river. "Big Tom" and I set up all night for about a week to protect our logs and equipment. We had ten-gauge shotguns and we'd shoot just over their heads. The way those guys were runnin', the bullets couldn't keep up with them. It became so rough the government sent in the National Guard to end the strike.

Big Tom was the son of Chief Thompson, of the Siletz Indian Tribe. He was a large man, six-foot-two-inches tall and he worked for Rex. He wore regular logger's clothing, no Indian beads, or feathers. He played the guitar in the evenings and gave Hazel a guitar so she could learn to play. She was sixteen that summer.

Rex's crew was logging northeast of Newport during 1934 and 1935. If you took the Coastal Highway north and turned in-land when nearing Otter Crest, you would find yourself in the area they were logging during those years. Usually they worked at the coast from mid spring to fall. It would rain hard in the spring and again in the fall. Since the ground was quite sandy at the coast, mud wasn't as much of an obstacle as it was around the Benton County area. The under brush at the coast was thick and difficult to deal with, much of it was about twelve-feet tall, and included salmon berry, devils club, huckle berry, salal and sword fern.

It was a job getting to the logs. One foggy day we'd high centered the cat in some brush and logs. Old Fred and I were out looking to see where to put the skid roads in and he couldn't find his way back to the cat. He wasn't far, only about one-hundred-feet away, he'd yell but I'd just keep the cat shut off so he couldn't figure out which way to go to get back to the cat. I could hear him out there, cussin' like mad. I got quite a kick out of that.

When Buck first came to work for us he would do anything we told him to, after a while he got plenty smart. Buck got so he could tell when we were about to play a trick on him.

Another time when we were laying out a logging road, Old Fred and I came out two miles from where we thought we were. We followed the railroad track back to where we should have been. Vip sure had a lot of fun over that. When we were cat logging at Siletz and Newport, Fred and Buck often set chokers for me. Fred was getting old by then and Rex would have both of them setting chokers for my cat. Fred was always getting Buck in trouble with the Yellow Jackets. Old Fred could see where those Yellow Jacket nests were from fifty yards away. Whenever he'd see a nest, he'd go to the "safe side of the log" and send Buck to the "Yellow Jacket side of the log." Those logs were often eight or nine feet in diameter so Fred would be fairly safe on the other side. Fred would yell; "What's the matter Buck? You having

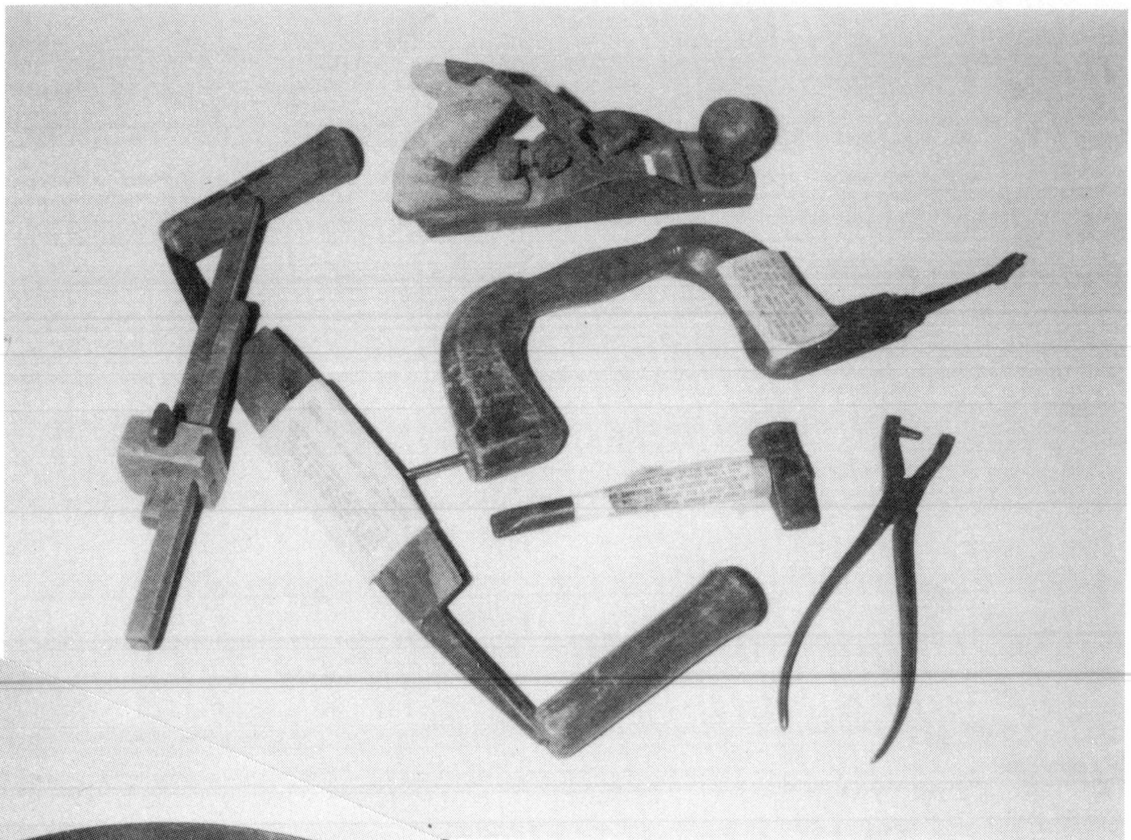

Some of Oliver Shroyer's hand tools. (TOP - LEFT-TO-RIGHT COUNTER-CLOCKWISE) **Carpenter's Graduated and Expansion Marking Gauge; Draw shave (draw knife) dated 1770; Hammer used to put lynch-pins into wagon wheel axles; Leather punch; Brace-and bit dated 1770; Small plane.** (CENTER) **Hand scythes for cutting tall grass.** (LOWER) **A hand grinder.**

George taking the Celtrac 40 through a shallow place in the Siletz River.

trouble getting that choker around that thing?" By that time those Yellow Jackets had found Buck. In the fall they would be real angry. Sometimes Buck would get so stung around his face he couldn't see out of his eyes. We'd have to let him set for two to three hours. Then Old Fred would have to work both sides of the log with the chokers. He didn't seem to mind, cause next time he'd see those Yellow Jackets, he'd herd Buck right there again. Sometimes Old Fred would get some up his pant's leg, then he'd run and cuss like the devil.

We stayed at different cabins Rex rented for us. I say different, because we would get too rough and get kicked out pretty often. One time I bought a string of fire crackers. You'd light one end and they'd go off in a series; I figured I could do something with 'em. One night after George McDowell, the guy we called "Kingfish," Al and Buck had gone to sleep I tossed these firecrackers under Kingfish's bed. It was sure funny to see Kingfish rise up-off-the-bed with each explosion. I could see him by the light from the firecrackers, and I could hear Kingfish go; Whoop! Whoop! Whoop! Right in time with each explosion. Although the bed set pretty high, it made holes in the under side of the mattress; I didn't think it would do that. When we came back from work the next night, we found our things outside. The owners had moved our things out, I can't blame them. Rex was pretty mad. He said; "I'll find you one more place, but this is the last time for that."

George's first truck, a K-7 International carrying an old growth 1-log load on the steep approach to the Siletz River high-water 200-foot-long bridge. See picture page 125 for the truck on the bridge.

The "high-water" bridge over the Siletz River.

In 1936 and 1937 we were logging along the Siletz river again. Rex and Ethel decided they wanted to start a cookhouse for us there at Siletz. Fred's wife, Lillian, was in charge of it. She fed us good, made good lunches and everything; she hired Mrs. Plunkett to help her. I remember she made the best darn chili in the whole world.

We were logging about three miles up the Siletz above Chief Thompson's place, at "Eddy Charlies." This was about one-third of the way up toward Logston. We logged close to the river; we swung the logs across the river on cables. The guy setting the tongs would ride them back and forth across the river. One real rainy day, the brakes on the yarder went out from being wet. Big Tom was ridding the tongs that day. Vip dropped him into the Siletz River from about twenty-feet up, he did that a couple of times. There was a huge bar of iron above him holding those tongs; the bar fell with him each time, that was pretty dangerous. That second time he came out; I told him he didn't look like a redman to me, he was "white as a sheet."

After we logged that all off, we had to log further in-land. It was too far to swing the logs across the river from there, so Rex had me build a "Low-Water Bridge." I was about thirty-years old then. That was in there, one mile above the cemetery, over there on Highway-B. Anytime Rex needed a bridge across a river, I'd oversee it.

We built a couple of bridges across the Siletz. The first bridge, built back in 1933, was a "High-Water Bridge," it was two-hundred-feet across. We built that when I was only about twenty-six-years old.

Rex gave me that job, I don't know why, guess he knew I'd kinda go ahead on my own. If we had to build a bridge when Fred wasn't there it was hard to get anything to build with; I'd have to go beg for cables and things. I'd go over to Camp 12, to get old line and stuff to use, that camp was built during World War I. Whenever Fred was there, he'd talk Rex into buying what we needed.

One time when I was taking the cat through the ford, a shallow crossing up from "Eddie Charlie's," a choker fell-off the cat. Big Tom jumped in the river and swam around there 'til he found that choker; he found that real quick, Tom was smart and not afraid to tackle any job.

Big Tom would take me across the Siletz river in a canoe. That was pretty scarry, I didn't swim. He'd just angle that canoe into the current and it would carry us right across the river. Tom was an excellent swimmer, he could get me out if we tipped over. He was a good man, a good worker. Later, he played football for Oregon Agricultural College now known as Oregon State University. *

* The name Oregon Agricultural College was officially changed to Oregon State College in 1937 and was reclassified by the Legislature as Oregon State University on March 6, 1961. (Editor)

CHAPTER 12.

LOGGING AT KINZUA, WINTER OF 1937

There was talk of Germany going to war against the various nations in Europe. The Japanese were beginning to pose a threat to the western United States and preparations were being made to avert any invasion attempts by them. Rumors were circulating, that the British had developed a light-weight fighter plane.

The threat of war and the poor overall economic picture was influencing the market for wood products. Demand was almost non-existent. Few homes were being built in the late Thirties. People had either used up their savings or were doing with what they had before the depression. No one knew what the future economic picture would be. It was the fall in the lumber market for wood that caused Rex to extend himself into the High Rolling Country of Eastern Oregon. This was the only work he could find for his men that winter. The sawmill at Kinzua (elevation 4,325 mountain summit) was buying a mixture of Douglas Fir, Ponderosa Pine, and Sugar Pine. They were producing tongue and grooved pine for a nearby manufacturing company that was building pine boxes, to serve as coffins.

Rex's men had set up their camp in Fossil, Oregon. (Elevation 2,654 ft.) This was about twenty miles from the slopes they would be logging that winter. Ethel had come along to help. She was devoted to Rex and his crew and had gained considerable experience over the years. She took care of the bookkeeping and ordered the supplies. She was always willing to listen with her heart, if someone needed to talk.

George had begun volunteering to do the high climbing. Rex never ask a man to do something dangerous. He always waited for someone to volunteer. George figured Rex didn't want to feel responsible for sending a guy up there. Well aware of the dangers of poor rigging, George was careful to properly secure it. When they were rigging-up, it always meant considerable time loss for the crew so, he worked as quickly as he could.

Rigging up a spar-tree for the yarding donkey was quite an undertaking. First, George identified a suitable standing tree, determining that it was sufficiently strong, being about four or five feet in diameter, and well located for their purposes. He began his ascent; cutting off all limbs then topping the tree at the desired height - 120 to 160 feet. George felt that it was safer to blast the tops out rather than cutting them out. After this had been accomplished, he would once again ascend the tree carrying in his shirt pocket a short stick wrapped with fishing line testing one-hundred pounds. After attaching one end of this fishing line to a spike in the tree he would tie a spike to the other end dropping it to the men below. They, in turn, tied a rope twice the length of the tree to this string, with the high climber retrieving it. Now they had something

sufficiently strong to support the pass-block's weight. Pass-blocks were normally six-by-one-and-a-half inches weighing about fifteen pounds.

The first block attached was the pass block. After securing it about three-or-four-feet down from the top and threading it with a small cable, they were ready to send up the first four-to-six plates that would hold the top-guy-lines. Each plate was about four-to-five-feet long, one-inch thick and four-to-five-inches wide, supporting three-to-four hooks needed for attaching the one-and-a-quarter-inch-diameter cables. Application of plates was done by placing them parallel to the tree and equal distances apart, and attaching them with rail-road-spikes through the holes in the plates. The railroad spikes used were somewhat like an oversized nail. The head of a spike was unique; it was flat on top, one-half-inch thick and off-center to the spike. Thus, flat against on one side of the spike, forming a cap off the opposite side, not round as the top of a nail is. The railroad-spikes were five-eighths-inch square, six-inches long and two sides come together at the tip, this made it difficult to hammer them into the tree. George usually carried a four-or-five-pound hammer to do it.

He then wrapped the top guy line cables once around the tree over the plates , then shackled them in place. The crew on the ground takes these cables out up to six-hundred feet then secures them to stumps. This provides stability to the spar tree.

Attachment was made by notching the stump with an ax, wrapping each cable around three-or-four times, and securing it with railroad-spikes. It was important that stumps below the spar-tree be deeply notched, as cables tended to slip up and off these stumps. If the cables slipped off, the spar-tree would topple breaking or destroying equipment, resulting in injury or loss of life. At best, a significant amount of lost time.

Next, the haulback-block weighing about seventy-five pounds with a fourteen-to-eighteen-inch sheave was attached. Placement was eight-to-ten-feet below the pass-block, allowing them to move it more easily while working to secure it. Afterward, threading the haul-back-block with the haywire, a somewhat flexible cable with a-quarter-to-five-sixteenths-inch diameter was necessary.

When they were using a spar-tree for yarding the logs, a high-lead-block was used. These high-lead-blocks ranged in size beginning with a sheave size of twenty-four-by-four-and-a-half inches, weighing about seven-hundred pounds to a sheave size of thirty-six-by-six-inches, weighing about fifteen-hundred pounds. Securing these proved to be a considerable task; exact placement was critical to the overall operation. As the logging progressed, moving this high-lead block around the tree put considerable strain on the man moving it; possibly more than the initial setup. The cable used with this block was two-to-three-thousand-feet long with about a one-and-one-eighth-inch diameter, weighing three-pounds or more per foot.

Finally, four-to-six belly-guy-lines kept the tree from buckling. These were attached two-thirds of the way up the spar-tree; by wrapping the cables once around the tree and shackling them in place. These cables radiated out to stumps only one-or-two-hundred-feet away. Attaching them required the crew to deeply notch these stumps, since they were closer and slippage was more likely.

At Kinzua they were cat logging, so the yarding was done with cats. George was setting up a hundred-foot spar-tree for loading the logs using a crotch-line. A smaller main-block and haul-back block was attached on each end of the spreader bar giving them double-power and more control.

One particularly cold day, George recalls becoming totally soaked with sweat from this strenuous work. George had finished this laborious task and had returned to the ground and was bending over a piece of equipment inspecting it, when a searing pain shot through his chest. George said; "It felt like someone stuck a knife in me." The guys didn't know what could have happened to him. Rex sent him home immediately since he wanted George to have his own doctor's care. Bill Adams and George Hoffman were heading down that afternoon so George rode along. The trip down to the valley was extremely hard on him. Helen found it quite frightening to see George doubled up with pain, and each breath was causing him great agony. When the doctor arrived, he diagnosed it quickly. George was suffering from a bout of pleurisy. The cold air had caused an inflammation in his lungs. The doctor explicitly directed him to stay in bed for the next couple of weeks. George was thankful for the pain medication the doctor gave him. He stayed flat on his back as he had been directed. This was one of the few times George willingly followed the doctor's orders to the letter. This was perhaps the most excruciating pain George would ever experience. After recuperating, he returned to work with the crew in the high country of Eastern Oregon.

This particularly blustery late November day the crew gathered together over a hot breakfast of ham and eggs, fried potatoes and the blackest of black coffee. There was the usual morning grumbling about such a day. They were just beginning to fully comprehend the dangers, such a day in the woods, could hold for them. They were use to the dreary winter rains in the Willamette Valley so these biting cold days were quite different. The equipment didn't work as well – always hard to start. They were a close knit crew and had worked together for many years. They'd shared in each other's troubles. Most of them were struggling with the additional loneliness exasperated by being away from their families. Not only during that past summer but now their winter work was taking them to a new environment. They were miles from their families, settled in the valley. One especially, felt this loneliness. Young Buck had finally taken a wife, Marie McDonald. She was use to living in Philomath and had decided to remain behind. Somewhat

Francis "Buck" Arbuckle with his new wife Marie McDonald

withdrawn into himself, Buck went through the motions of leaving the bunkhouse to depart to the wind-blown slopes where they were logging.

Buck climbed into the pickup with the rest of the guys. Little did anyone suspect that before the day was over one of their greatest fears would become a living reality. Their memories would, as long as they lived, fill them with unutterable pain. Their senses would clearly etch this day's experience into their minds. It would seem as if it were only yesterday; holding in its grip the pain of this day forever.

Buck had, over the years become a seasoned logger and he was operating a cat, as was George. George had taught Buck how to do it and it came to him naturally. Mechanically inclined, Buck moved easily into his role as cat operator. The climate in the higher elevations is quite different than that in the valley and along the coast where they were use to working. The ground up here was very dry and these pines had shallow-root systems. The crew was not yet use to the conditions specific to logging this particular area. Most of the remaining trees were standing on the slope above them. They were experiencing severe down drafts on this slope. The crew had decided that perhaps they should have logged above first, leaving the uncut trees on the lower part of the slope.

Buck was working to remove a log that had hung up on a stump. To get the log past that stump he had to adjust the choker, this meant he had to leave the cat running while changing the choker. He'd just finished adjusting it and was returning to his cat when a tremendous gust of wind wrenched a tree out of the ground. He was in somewhat of a draw, below the trees. The men witnessing this cruel act of nature yelled to alert Buck to the danger. He continued on, oblivious to the eminent tragedy. Orville and others saw it coming and yelled but Buck seemed not to hear them at all. They believe the noise from the cat engine kept Buck from hearing the great cracking of the roots as the tree broke loose from the soil. Their attempt to alert him to the danger went unnoticed. This tree, unprotected from the wind, couldn't stand the brunt of it.

George was on a cat near by. He could tell something had gone terribly wrong. He could see the tree as it fell. He jumped from his cat and ran toward the area Buck was working in. He arrived just as the falling tree settled down for the last time. The men witnessing the tragedy said the tree came down and hit Buck in the back, bounced up and settled back down. George said the part of the tree that hit Buck looked a bit larger than a telephone pole. It had about a twenty-four inch diameter on the butt and was approximately one-hundred-fifteen-feet tall. The crew had to pack Buck out of the woods on a stretcher to get him up to the pickup. It was an extremely cold day, and the men had great difficulty getting the truck started. Once they made it back to Fossil, Rex drove and Ethel held Buck in her arms. Ethel would need to deal with the hospital administrative details, call Buck's family and his new wife, Marie.

Buck died along the highway, at Celilo, below the falls where the Indians have their fishing grounds. Today the waters from The Dalles Dam covers most of the area. He was put to rest at Oak Lawn Cemetery, here in the valley, on November 22, 1937. Frances Eugene Arbuckle. Born 1914—Died 1937. He'd lived only a short, twenty-three years before he died. A bronze plaque marks his grave.

Buck could sing beautiful Irish tunes, and this brought the men close to him. They all thought a great deal of this young man; he'd worked seven years with them, before he died. George had come to think of him as a younger brother.

Helen looked up to see George filling the doorway with his large frame. She was surprised to see him. One glance and his eyes told her of his terrible pain. No doubt, there had been a tragedy that day, he'd come home unexpectedly. Her own fear of loosing George had not materialized. George was smiling softly at the children that ran to greet him. He held the children and played with them until bedtime. He seemed so glad to see her, so happy to be home. That night, he cried all night. Morning light found his pillow, and the sheets on his bed completely soaked.

Rex's crew returned to the slopes, this tragedy leaving them stunned and lonely. The Christmas season, overshadowed by the sadness surrounding Buck's death, was not a cheerful time, that year. The New Year brought no release, the cold only deepened and the winter dragged on mercilessly. Returning to the valley late March of 1938 resulted in alleviating some of their

pain. It was good to be home, to leave the reminders of nature's cruel act behind.

April 1938, the economy was beginning to stabilize. Rex was expanding his operations; there would be two crews this year, and they would need additional equipment. He bought a yarder that spring from another logging company they'd been using it over there near Arlie, between there and Pedee. Vip and George were going to move this yarder down to the coast. It was on a sled thirty-six-feet long and only six-feet wide.

> **Vip and I were moving the yarder on a Sunday. We were going to move it from a rollway onto a truck and trailer. The rollway was much higher than the bed of the truck. Vip was in charge; but two different times, I said; "Vip, don't you think we should snub it onto a stump?" He said; "No! It will be OK."**

The sled with the yarder on it weighed five-to-six-thousand pounds. The sled held the yarder, two drums and an engine. Adding to its precariousness, was a high roof covering the hoist, or engine and drums, on the sled, and it set on a high rollway. Vip climbed onto it so he could use the drums to move it, he was going to move it by pulling both the haulback and main-line at the same time since they were fastened to each end of the sled. He expected it to move forward, onto the trailer.

Lowering the sled, caused it to set at a forty-five-degree angle. The rollway was too high for the truck bed; the sled hit level, tipping on over momentarily. The sled flipped over the truck and trailer bunks throwing Vip to the ground where earlier, they'd left some timbers. The sled and yarder came crashing down pinning Vip with one leg bent up to his chest in only six inches of space. Those timbers kept the sled from killing Vip. Grabbing the only remaining timber and using it as a lever, George lifted the sled-off Vip. This timber, being twenty-four-feet long provided the mechanical advantage George needed. By using his weight as leverage, and working his way toward the end of the timber he brought it down low enough that his feet could touch the ground. Next, noticing a cord of wood stacked near by he managed to carefully re-stack this around the end of the timber while remaining on it. He stacked enough of this wood on top of the timber to equal his own weight plus an additional one-hundred-fifty pounds. This held the sled securely in this raised position, assuring that it would not move, enabling him to pull Vip to safety,

> **I didn't have much time, not much to work with, Vip was unconscious. He "came-to" a couple-of-times. He handed me his billfold and said; "George I want you to give this to Pearl -- I may not make it." I took it, I didn't want to argue with him. First I got him out, then I took the seats out of the truck, laid him on them and covered him with coats. I stopped at a close farm house and asked this guy to go stay with Vip. He told me he was just back from church and didn't do**

> anything on Sundays. I was always going back to whip that guy. Then I drove ten miles to the nearest phone. It took that ambulance an hour and a-half to come from Corvallis to get Vip.

He had internal injuries affecting his liver and spleen, also breaking his hip and leg in several places. Vip had come close to loosing his life in this accident. He recovered a year later then he worked for many years after that.

CHAPTER 13

LOGGING FROM 1938 THROUGH 1945

Early spring of 1938 found George in his shop there in Gellatly Canyon, repairing Rex's equipment. The cold days on the slopes in Eastern Oregon had been hard on their equipment and trucks. The crew were not accustom to the cold and it had been an exceptionally difficult winter. George's own fight against pleurisy had taken its toll and Buck's death had made the winter seem long and difficult. George was happy to be spending more time near home these days. He found Helen's presence reassuring and the laughter of his young daughters clearly brought him happiness.

> **A few days later, Rex came over to see how it was going. I had three of his cats torn apart and I was overhauling them. I had parts scattered all over that little shop, pistons all over the darn place. He came in, looked around and he said; "How you ever going to get these things back together? I said; 'I'm not. I quit! I don't know where they go.' I started walking toward the door. Rex said; 'Don't you dare quit I'll shoot you. You go back in there and put them together, then you can quit.'**
>
> **Rex bought some diesel cats earlier that year, it was the first diesels we'd had. The cats we'd had before, had gasoline engines and they used spark plugs. Rex was going up to Portland for parts and I asked him to pick up some spark plugs for the diesel cats. When Rex came back from Portland, I asked him about the spark plugs. He says; 'Shroyer you can't pull that on me. I know diesels don't have spark plugs.'**
>
> **Later on I was up there in Portland with Rex, getting parts. The guy's in there got to laughin' about Rex ordering those spark plugs for the diesel cats. Rex said; 'I knew all the time, I never did order those spark plugs.' Those guys went in there, brought out that order and showed it to us. It was the order he'd written, for those spark plugs. He'd been telling me all that time that he'd known diesels didn't need spark plugs. I was always getting him to do something crazy like that.**

Another time I had alot of fun with Rex was when I asked to be paid for a twenty-six hour day. By me working long hours we were able to save quite a bit of money. I was the only one who did that. If anything needed done, why, I'd do it -- I'd work every Saturday and Sunday. Sometimes I had a half-day off on Sunday. I had Christmas day off, that's about all. So one time, I took some hours off other days so I could have a twenty-six hour day. I figured Rex would jump me for that, he never did. Old Fred told me later that when Ethel mentioned it, Rex said; "Don't say a thing, Shroyer's just doing that so I'll get on him. Just ignore it and pay him for his twenty-six-hour day. He's always playing tricks on me. I sure would like to can that son of a gun, but I can't do without him.

I remember playing a pretty good trick on Grandpa one time. In the winter time, we kept our bull tied up, kinda under one side of that shop. Grandpa would take him down to the creek that ran through our place so he could get some water each day.

That young bull would lumber along behind Grandpa real slow. I was working in the shop one day when I heard Grandpa out there cussin' that bull; he wouldn't lead, he wanted to stay and eat his grain. So, I just picked up my BB-Gun and shot that bull right in the butt. That bull, he really took-off. Grandpa could barely keep up with him. For a minute, I thought I might have to shoot Grandpa so he could keep up with the bull!

The summer of 1938 Rex started logging north of Newport. Iva Edwards and Marvelle Tatom started a boarding house for us in Newport. I stayed there that summer, it cost a dollar and a quarter a day for room and board. There was a pretty lively crew at the "Boil House," with people playing jokes on each other all the time. They had a real nice skating rink in Newport. I had so much fun skating there, I bought Joyce and Lucy some skates so they could skate in our basement at home.

On Thursday night, almost everybody who worked for Rex went skating, especially the men. The women didn't go as often, when they did, the men always came home earlier than the women. This one night, we all went skating.

In the "Boil House," the women stayed downstairs where they had a bathroom. They had just a little light in it; they had a string to pull to turn it on. Cal Tatom, Orville and I made up a dummy earlier that day. Cal made a head for it and we put Orville's old clothes on it.

When we came home from skating at midnight, we put it on the toilet. We knew the girls would go in the bathroom when they came home. We'd gone to bed, but we didn't go to sleep cause we knew what was going to happen. The girls came in pretty soon, opened the bathroom door and turned on the light. There was that guy settin' on the toilet, it scared them to death. They screamed! You could hear them all over town; we were trying to be quiet and not wake anybody up, it was pretty late. Next morning the girls moved the dummy to a chair by the kitchen table. That weekend when Rex came by, he looked over at the dummy and said; 'I see Orville's back from hunting all ready.'

There was this old guy that ran the service station at Newport. Everyone called him "Dad," he was maybe sixty-five. He told me he'd worked for McCormick-Deering, a big tractor outfit. He said that they gave him a physical when he retired. The doctor told him; 'If you find something to do, to keep busy; you'll probably live fifteen years or more. If you just go home and sit in a rocking chair, you'll probably live only a couple of years.' I was surprised it made that much difference. Well, that old guy got that service station running and kept pretty busy, last I knew he'd been there twelve or fifteen years.

Anyway this old guy, he lived right there at the station. One day he had a big kettle of stew on the stove and I asked him what he put in it. He said; 'Everything, the more you put in the better it is.' I said: 'Is that right"? He said: 'Oh yes'! Then he went out to see what someone wanted. Right above the old stove he had a place where he kept some of his tools, wires, screw drivers and stuff. I took an old pair of pliers, rustier than the devil and an old Crescent wrench and put them in his stew. So, next day I asked him; "How'd you like your stew. He said; 'Oh. Just fine.' Never let on like there was anything wrong with it.

One morning, Al, Kingfish, and I stopped there for some gas; we were in Old Al's car. That old guy had a cup of coffee in there that he was drinking, it was about two-thirds full. While he was filling up the truck, I went in there to look at the ice cream because we usually got an ice cream bar apiece. I saw a salt shaker settin' there, so I put alot of that in his coffee. He came in pretty quick. The minute he tasted it he knew something was wrong; he just gulped the rest of it right down. I had no idea he would drink it. He didn't want to let on that anything had happened.

Next few days, Rex stopped there. That old guy, told him; 'Your men have been playing some pretty tough jokes on me. I've got to do something to get even.' Not long after that, we stopped there one morning. He said; 'Well my electricity went off last night and I got some ice cream that's melting.' He had three of them lying up there on top. 'You guys better each take a quart.' Each of us took one, opened it up and started eating it. I thought mine felt kind of soft on the inside when I was eating it. I thought; 'Boy! That's kind of unusual.' It should have thawed on the outside. So, I dug down in it -- there was horse manure in it. I told the other guys I didn't believe that I needed to eat any more of mine. Old Kingfish said: 'Well, I'll eat it.' So, I handed it to him. He dug in there and ate a couple spoons full before he saw the horse manure in the bottom. That old man got even with us all right, we never mentioned it, he didn't either.

Winter found Rex's crew closer to home; he had a contract for logging Douglas Fir out of the Honey Grove area. Honey Grove is a pocket of rain drenched alternately set hills lying in a valley southeast of Alsea. Flanking this on the east is a ridge known as the "Railroad Grade" rising up a thousand feet out of the earth and running north and south for about twenty miles. Today, there is a small garlic farm lying in the narrow valley along the creek. The enormous growth of the flowers springing up from the soil attests to it's richness. The second growth timber is thick, and will, in a few years provide a harvestable supply of Douglas Fir.

Rex had a crew of us in there working with some dynamite one time. We were blasting out some big stumps and some rock that was in our way. We were hauling a big sled of powder in. Carl Bennett was running the cat, my brother was riding in the second seat up front and I was settin' back there on the sled with the powder. Old Carl would look back ever once in a while, I'd be making up some dynamite primers. Soon as he'd turn around I'd work on a dummy one. Gail saw what I was doing. Old Carl looked back and I made like I was lighting one of those; I threw that up in the seat with him. I tell you, he jumped off that cat. Never switched the cat off, took it out of gear or nothing. He took off running for all he was worth; just then, I started laughin'. I'm sure Carl will never forget that as long as he lives.

On cold winter days, when snow is on the ground, sometimes it's hard to get the loading donkey to fire up. One morning, Vip was helping to start it. He was standing near so he could pour gasoline into the carburetor. That old loading donkey would cough, sputter and spit out flaming gasoline. Since Vip was standing so close, the flames fell on him catching his "tin pants" afire. Boy! did he take-off running; that only fanned the fire and caused it to burn more. Orville was working close by, so he took-off after Vip, tackled him and rolled him in the snow to put the fire out. Before Orville knew it there went Vip again—running like the devil. Orville tackled him a second time and rolled him in the snow. As Vip was getting up and Orville was still trying to catch his breath from chasing after him. Orville says; "Next time you catch yourself afire, you better run towards me not away from me. His hair and eyebrows were singed, there were some burns around his neck and it burned his hands some. That snow worked good to cool the pain from his burns. Vip was able to work the rest of

the day, it didn't seem to bother him. We got that donkey started without Vip getting himself afire again.

Rex asked me to make him a new landing over at Honey Grove. That was 1942. I'd been driving from the coast over to Honey Grove to get that work done for him. Earl Roberts and Kenny Stanton were working as choker setters, Orville was the hook tender and Kingfish was the donkey puncher. He was running that diesel yarder, right after lunch, they were bringing in a big log. The guy's were all watching that log being yarded in, Kenny was leaning on a four-foot stump, Earl and Orville were standing about three-feet away, on his right. Orville yelled for Kingfish to tighten the line. The haul-back line picked up a limb that was ten-feet long with only about a three-inch diameter, wasn't very big. Kenny said later, that he felt the wind as that limb came whizzing by. It hit Earl on the side of his head and split that cotton cap he was wearing. Orville sent in a signal of whistle blasts for man injured. I looked down below, there was Earl lying on the ground; I was up on a ridge near by picking out the new landing when I heard the whistles. Hard hats were worn beginning in the late Sixties.

Working in the woods was mighty dangerous; you never knew when it was going to be 'your time to die'. I'd been standing there with them just minutes before. We carried him out—we were all stunned. Couldn't believe it could happen that quick. We got him to Old Doc Anderson soon as we could, he took one look at Earl and told Rex; "I don't think I can save this one." Earl died later that afternoon leaving behind a wife and a young family. He was only thirty-one.

While George was working at Honey Grove, he normally arose before dawn each morning, fed himself in fifteen minutes and drove twenty miles to the foot of Alsea Mountain. Arriving at the worksite in the first light of day before the early morning winter fog began to lift. George drove to work each morning in his Plymouth pickup. In those days the men had to buy their own pickups. Early one morning, Rex called to ask George to pick up a couple of the crew, this put him behind schedule. George was following a school bus as they traveled toward Honey Grove. Much to his surprise, the empty school bus turned sharply to the left directly in front of him. He was unable to stop and broad sided the bus. The driver jumped out shouting; 'What did you do that for"? Everybody knows I turn around here. George's Plymouth pickup was totaled, but none of the men were badly injured. Only one of them had any visible injury was Old Al. It was a week or two before the yellow-blue coloration across his forehead and around his eyes faded and the natural healthy glow of an outdoor man returned. George bought a new Dodge pickup to replace the totaled Plymouth as he needed something to drive back and forth to work.

George requested that Helen speak with the judge on his behalf as he was much too busy to go to court. Helen told the judge how the bus driver had jumped out demanding to know why George had broadsided him -- when everyone knew that was where the school bus turned around. Helen told the judge; 'The school bus driver suddenly turned across the road, directly in front of George without signaling.' She explained that George previously, had always traveled that road before six o'clock in the morning. George would have no knowledge regarding the school bus's morning routine. The judge ruled in George's favor and against the school bus driver for not signaling before turning.

Rex had a crew working on the railroad grade beginning in 1939. In the early days the railroad started out at Corvallis, went south for about ten miles then turned west, went past Hull's sawmill and on over to Dawson. From Dawson it went up onto the ridge known as the Railroad Grade.

> When there were trestles to build or railroad grades to work on they brought in the Chinamen in the early days. There were lots of them working here, they were hard workers and they lived right there, in the work trains parked on railroad sidings. When I was just a kid, I remember one of those was parked on a siding there at Summit, there were lots of Chinamen living there. Us kids would make fun of them, we'd go by and point at them. One Old Chinaman got mad at us about that. When he came down where us kids were, he had a knife that was about three feet long. I don't think he intended to hurt us; just meant to scare us. He sure did! We ran, I'll tell you. I bet that guy is laughin' yet, at how we took-off.
>
> The Corvallis Lumber Company logged the Railroad Grade area until 1939. Then, they decided not to use the railroad any longer so they took up all the rails and railroad ties. The Old Railroad Grade became a logging road. Corvallis Lumber Company decided to contract the logging, they offered that to Rex. There were lots of spikes up there, those would ruin a tire. If a loaded log truck ran over a spike, that could cause a bad accident. Rex paid Bob Bennett's sons a dime a spike to pick those up. I remember watching Dick and Ray Bennett walk up and down the old Railroad Grade behind Ralph Gee's grader picking up those spikes. We hauled those logs out on that road for several years after that. Mama and I were living at Gellatly Canyon when we were working up there on the Railroad Grade. I would drive to Philomath, then take the Bellfountain Road out about fifteen miles and go west for five more miles to get there.

I remember the tallest tree I ever climbed was up there on the Railroad Grade. It was a two-hundred-foot spar-tree Corvallis Lumber Company had used.

It looked like they had rigged it up for loading logs out of there. We were needing a pass-block and Rex was watching his money, he didn't want to buy one. I'd noticed one up in this big-old-spar-tree. I figured I'd climb it and get that block out of there, we sure needed it. After work one night, when everyone else had gone home, I started up that tree. There was a wide band, about twenty feet up, where the tree was completely bare-of-bark. That kind of tree is pretty dangerous after it becomes old and dry, cause the only thing holding that thick bark on there is the small six-inch stubs left when they cut the branches off it. Well, I should have known better but I'd decided I was going up there. I'd made it up about forty feet when that twenty foot section of bark came loose from the tree and dropped with me on it. Boy! That was scary. After it stopped falling with me, I was able to free myself and I went on up the tree. When I got to the top I found that block wasn't any good, after all that. I could have hurt myself real bad, I was all alone and I could have hung up there all night. That was the dumbest thing -- I ever did.

About that time, when I climbed a real tall tree, I was starting to run out of air. I sure couldn't figure that out; that was about the time we got to living on strawberries. Mama and Grandpa had made a nice big patch of the Marshall variety of strawberries down there across the creek; just out the kitchen window. Those were the best strawberries I'd ever seen, they grew as high as my knee, that's pretty high for a strawberry plant. You'd see hundreds of those little green berries on them, I can still remember those strawberries. I was allergic to them; I'd break out in hives all over wherever I'd eat them. Boy! I'd itch. That must have been what started my asthma. I did quite a bit of high-climbing in those days; I'd get up there about thirty feet and be clear out of air. I couldn't imagine what in the devil was the matter with me. Finally Doctor Whitelaw told me I was having asthma.

Old Doc Hamm worked with us up there on the Railroad Grade, he always brought lots of boiled eggs with him for dinner. He'd take two eggs at a time and crack them hard over his forehead. One time, I figured I would cure him of that; so, I just changed them with raw ones. We were setting around the fire eating dinner. Good Lord were we surprised. Egg just run down the whole front of him. Bruce Hamm, Doc's brother, pret'near died laughing over that. Doc tried to blame it on me, but I told him; 'I would never do a thing like that'! Lorene his wife was good natured so, I said; 'Lorene must have done that.'

We called Hal Hamm, "Doc," because he believed in the theory a guy named Doc Townsen [*sic.*] had about rebuilding the economy after the depression. He believed the government should give everyone two-hundred dollars a month.* Then, everyone was supposed to spend their money each month; that way they would just trade their money around. They tried it in some small towns, but some of those guys saved all their money; so it didn't work.

This particular winter, Rex had more than the usual number of men working for him, it was just about Christmas time. The depression was just beginning to let up some and Rex was planning on expanding his operations to two, possibly three, winter sites rather than one. His men would be working both in Honey Grove and on the railroad-grade. They were trying to get as much out of their equipment as possible. Early morning, found George out in the woods starting the loading donkey's engine, Gail was helping him. Gail had recently begun working for Rex again and George was glad to see more of him. That morning Bob Bennett had arrived early -- he was ecstatic. He and his wife had just had a baby girl, "Bobbie" now he had a namesake. Bob and his wife had thirteen children, Bobbie being their last.

Old Bob went over to Gail and offered him a cigar. Gail said; "What's that for? Bob said; ‘Well, I had a baby girl and I'm celebrating.’ Gail said; ‘When the Good Lord said 'Go forth and replenish the earth' He didn't mean for one 'so-and-so' to do it all.’ Bob was in the course of changing from a good logger to a good Christian and he was trying to be good, but he could cuss. He was the best cusser I ever heard; people would get him mad, just to hear him cuss. Well, he took out after Gail and as he ran past his truck he grabbed his ax. Gail was running down the road and ‘laughin' so’ -- he could hardly run. Old Bob was running after him, swinging that ax he was carrying. About then, Old Rex came up and he asked; ‘What's the matter with those crazy guys? I saw them running down the road as I came in, Bob right after Gail.’ When I told him, he got quite a kick out of that.

Bob was head trucker and he wanted all their trucks to be operating efficiently. He took real good care of his truck, checked the water, oil and kept air in the tires. He didn't have much trouble with his truck. It usually lasted twice as long as the other guy's. He got a truck every four-years; the others got one every two-years. Bob liked Chevrolets or Internationals, he didn't like to drive Fords.

George and Helen were delighted over the birth of their first son, George Edward Shroyer, II. Now George would have someone to follow in his footsteps. "Georgie" would grow to be a young man full of the understanding and knowledge acquired by working in the woods. At age four, he could already sense his father's mechanical capacity.

*Francis Everett Townsend (1867-1960), a physician. He was the originator and head of the Townsend Recovery Plan announced January 1, 1934. This was in the depths of the Great Depression. His plan was to give citizens over 60 years of age pensions of $200 a month to be paid for by a national sales tax.

At that time, George was working in the Klickitat Area, which is northwest of Alsea and west of Mary's Peak, running up towards Harlan. George built eighteen miles of road there opening the area from Alsea.

> **When Georgie was a baby, I was working up there, building the Klickatat Road. One evening, when I was driving the cat back down, Old Buryl Tom was riding in the seat beside me; the old cats had a "double seat" in them. We were coming down through the place where Bethel Vernon lived. He was a-half brother to Buryl. I remember there was half-a-dozen Plymouth Rocks, they were out in the road running ahead of me. The road was just rocks in the middle and I didn't dare tear it up because Bethel Vernon and his wife drove on it.**
>
> **Three or four of those chickens were fifty-feet ahead of the cat. Old Buryl, just pulled out his ole-22 revolver that he carried with him all the time. He said; 'I'm gonna have chicken for supper.' He just bang! bang! and shot two of those chickens just like that. Boy! This Old Lady was pretty mean; she had Rex and I scared to death. So I said; 'You're gonna get yourself killed.' He just said, 'Nope, I'm gonna have fried chicken for supper.' He got off the cat, went over, picked up those two chickens and started walking for that house.**
>
> **Boy, I didn't want any of that so I got out of there. Anybody that would up-and shoot his brother's chickens; then have nerve enough to go to the house and tell them he wants fried chicken for supper -- that's exactly what he did. I told Rex about it and he thought that was pretty funny. Rex told me he sure wouldn't want to be around there, when that happened.**
>
> **That Klickitat Road had lots of areas that were a real mucky, real soupy and difficult to work with. One time we had a truck with a load of logs, it got stuck in one of those places. We had to put a D-8 cat in front and one behind so we could pull it out of there. In those days, we built 'Corduroy Roads' in those swampy places. We were using logs that were eight to ten feet long and a foot or more wide laid side-by-side next to each other. After we put those in, there was still about one-and-a-half feet of muck on them. The guys couldn't figure out what we were going to do about that. We decided to haul in one or two feet of rock for it. Gail was hauling that in for us. We use to call that 'Clemens' Rock', it was about two-thirds dirt and one-third rock. We'd put that in there, but the mud was still coming up through the road. Slim Nutter and I dug a trench along the side to let the mud run out. Rex needed those logs out of there, he had a payment coming up.**
>
> **We had a landing near there with a little spring near by. We had a four-inch irrigation pipe stuck in there. Finally, I figured that spring**

would give me plenty of water so I back-bladed that section of road. There were layers of dirt and rock, I was washing the dirt out of the rock. There was more water on the inside rut so I went to slushing that mud up over the outside edge. I finally washed that mud out of the rock, that made a real solid road bed in there. Those guy's couldn't figure out why I was slushing that water over the other side. If you clean that mud out of the rock then all you have left is rock and that will stay where you put it.

One time out behind old Wanda Hayden's place, when Old Bud Reynolds and I were building that road, we had to take our powder and diesel up a steep hill. We didn't have the road all finished, there was about a mile and a half in there that we were working on. We had to get up above where we wanted the road, to work the fill down hill. We had to go in any way we could, had to go through some fields. There were two horses in that field, but this one horse, he just enjoyed us. Every night, that horse would wait to see what we were going to do. He'd follow us up to the top of that hill, then we would roll an empty barrel down at him or tie one onto his tail. An empty barrel will jump about thirty feet sometimes and that horse would just run. Take-off and just have the time of his life. Whistled at both ends. Sometimes that horse looked like he was laughin'. When we wired a barrel to his tail, he'd run and kick up his heels, it'd fall-off about a quarter of a mile away. That was down the hill below Hayden's place on the 'Hundred Years Ranch'.

That Klickitat Road went up there by the lake, it was too far to go home every night. Mrs. Grant lived up there, she had a big house so she kind-of-adopted us. She use to feed us and let us sleep upstairs in some of her bedrooms. She had a player piano, I sure liked that, got so I could play real good! When it came time for supper, I remember seeing Ernie Solie, another of Rex's cat drivers, looking out the window. He'd stand at the window and count the goats before he'd eat. Ernie said; 'I'll eat anything else, but I don't want to eat any-old-goat.'

There was this old guy up there, he'd married a young girl half his age. She'd lived up there most all of her life, she hadn't seen many men before she married. His name was Ed Sharp and his wife's name was Honey. He was a sharpshooter and had only one eye. He was darn good. One time I asked him; "Just how good are you?" He pulled his gun out and shot six pine cones off a tree nearby, never missed or stopped once. Some of us guy's got to going by their place and flashing our lights in his windows at night. He just grabbed his gun and let off several shots in our direction. That put an end to that. We figured we wouldn't live long if we kept that up. Honey was sweet on Rex, so she'd come up there whenever she got a chance. That was mighty dangerous since her husband was so good with a gun and didn't mind using it. He was a mean old cuss. Rex never hung around long when he came up to check on things. He was always glad to get out of there.

That next winter, we were working up on Prairie Mountain. Often in the cold weather the landing crew kept a fire going, it was kinda nice to have a fire to warm up by and to eat by in the winter. The chaser always had a fire going on the landing, that way we could clean up a few of the branches and some of the bark. We had two feet of snow on the ground that winter. When we got to work one morning we saw some smoke circling up out of the snow, we thought that was kinda funny. The night before, the fire hadn't been put completely out. It got to smoldering under the snow and worked its way out into some slash we had near there. Next thing we knew the fire was in a large stack of logs. The fire was moving under the snow. Once that fire got started we couldn't get it out, it snowed and burned for two or three weeks. The darnedest thing you ever saw—a puff of smoke coming out from under the snow. That was one of the worse fires we ever had. Rex got pretty excited about that, those were his logs.

Loading with a single-stick boom on Eckman's Slough. Seated on log: (INSET: LEFT-TO-RIGHT) Galen "Red" Stanturf, George Shroyer, Vip Gee. On cat: Ernie Solie

CHAPTER 14

LOGGING IN THE EARLY FORTIES – THE WALDPORT VENTURE

Nineteen-forty-one was a difficult year for the Shroyers. Late in May, George and Helen suffered the cruel loss of a baby girl, the baby living for only one day yet never to be forgotten. Helen had contracted German Measles during her pregnancy. This sadness surrounds them yet, as they recall with sorrow the loss of this tiny one.

As the year progressed, it seemed to be unusually harsh. While still saddened by their loss, George experienced his first asthma attack. The marvelous strawberries they so looked forward to in the spring triggered it. This tendency was to follow him throughout his life resulting in an ever intensifying struggle.

Rex promoted George into the position of "Superintendent of Clemens Logging," giving him increased responsibility. One Sunday afternoon, George was returning from Eckman's Slough after his weekend labor along the coast when he heard the news of the bombing of Pearl Harbor on his pickup's radio. This would change the lives of many Americans. The United States declared War on Japan the next day, December 8, 1941, ending the year in the far reaching tragedy of a world at war.

> **Rex had government contracts that kept us logging spruce at the coast during the early Forties. The British Government was buying spruce in those days to make their new fighter planes. Spruce was a lightweight wood, but it was real tough. They were using it for wings and fuselages.* The spruce trees at the coast were huge, those logs were so big that a man couldn't reach across them. One spruce had a thirteen-foot diameter; we didn't fall it because we couldn't get a permit to haul it down the road, so we left it. We logged nothing but spruce for a couple of years.**
>
> **The first year or more we'd drive back and forth to the coast several times a week, sometimes I stayed over at a hotel and ate at a little restaurant. We were logging out of Eckman's Slough, it was two or three miles inland from the coast. There was only one guy we ever had who didn't like any joking going on, that was Old Bruce Hamm. He didn't want anyone interfering; he wanted to work. If he wanted to play, he'd just take a day off to play.**
>
> **When we stayed over, we ate at this little restaurant. That night there was Ernie, Bruce and a couple other guys. Ernie and I had put a snake in Bruce's dinner bucket. They had an eating place kind of out there at the north end of Waldport. You could get crab legs with the shell off them, you know. You could get a heck of a good supper, all you could eat of those crab legs. The women would be cleaning up our dinner**

* The spruce was for the deHaviland "Mosquito" fighter/bomber/reconnaissance airplane of which over 7,000 were built. It proved to be one of Britain's most successful wartime aircraft. (Editor)

buckets while we were eating. This old lady and her two daughters opened this dinner bucket to clean it. Well, there was this old snake in there. One of them, let out a yell and dropped Bruce's bucket. She came hopping out there with Bruce's bucket. She should have known it wasn't Bruce that put it in there—you don't put a snake in your own bucket. She jumped all over him; then she wouldn't give him any more pie, he never got any dessert after that. Old Bruce was the last one to do something like that, Bruce was a thin guy and he didn't gain any weight on that job!

A few years earlier, the government had tried to have service men log the spruce used in building airplanes. They had one-hundred-twenty-eight men working for about eight months to get out-half-a million board feet of spruce. After that, the Government decided to contract the logging. Rex's crew of about twenty men could log out more than that in a couple of months. So, the government designated logging as a 'Triple-A-Industry.' We were as essential as the soldiers, they needed us to keep logging out the spruce for their airplanes.

Rex had some of those contracts. He and Ethel had to keep getting deferments for about a dozen or more of us to keep us out of the service so we could keep logging. The men over here under that regulation should have been in the service, but getting the spruce was as important to the war effort than us being shot at over there, that kept Ernie out. We couldn't get the logs out without him. Rex had to make a report out for Ernie every week. He'd work any time, day or night, seven days a week in those days. The minute the war was over, Ernie said: 'No more Saturday or Sunday work'!

Helen said; Joyce was ten years old then and she was very ill during the fall of 1941. She had a kidney infection that put her in the hospital for a time then she was sick for six months after that. We were afraid we were going to lose her. They didn't have any antibiotics for people to take in those days and kidney infections were pretty serious. George was working for Rex, mostly at the coast. He'd come home to see how Joyce was whenever he had the chance. It was such a long drive each way to the coast and back.

George said:

During World War II, times were pretty tough. When they started rationing the gas, you couldn't get enough to drive back and forth to the coast. I'd been doing that for more than a year when that happened.

Mama and I decided to move closer to my work. On moving day, we took two vehicles. I drove the pickup and trailer and Mama drove ahead of me in the car. We made two trips down to the coast. First we loaded up the piano, beds, clothes and kitchen things. On our last trip Lucy rode with me, we took the milk cow and the rest of the furniture. Grandpa and Joyce went with Mama. They helped hold the house plants and take care of Georgie. I remember Mama had a flat tire along the way that time. Rationing of sugar, tires and things like that had started, so our tires were getting pretty worn out.

We moved into Waldport in November of 1942, and stayed there until November of 1943. World War II was about half-way over by then. Once we moved to the coast, Rex Clemens had us logging from Eckman's Slough on over to Yachats. Some of the time we even logged down there in the Blodgett track, below Yachats.

Living at the coast was kinda scary for the kids. Some Japanese navy vessels had come pretty close to the coast and it was feared that they might attack at any time.* There were black-out restrictions while the family lived there. These included using black-outs over their headlights. These were heavy pieces of tape placed over their headlights. They had small slits in them, allowing only a little light to shine down toward the ground. Also, since any light on a rise might be seen up to nineteen miles out across the ocean, the houses along the coast had special blinds. The blinds covering the windows, were dark green canvas allowing no light to shine out during the night.

Life at the coast was an exciting time for the children who had lived relatively sheltered lives prior to the move. Joyce was in sixth grade and Lucy was in fifth grade attending school in Waldport that year. Their little brother Georgie was only four years old so he stayed home with Helen and Mr. Hunter. At first, the family lived in a house that faced the Pacific Ocean, not far from Stump Beach. There, the winter gales filled the nights with the thunderous sound of crashing waves. Helen wanted a place in town so the girls could walk to school. Soon they moved into a small house in Old Waldport. This house holds great memories of the days they lived in a small town. It was located right where the supermarket is today.

The living room was the warmest place in the house. The family would all sit around the fire in the evenings before bed-time reading. Nancy Drew Books were among their favorites in those years. Joyce said; "I can still remember Daddy saying; "I sure wish we had some carrots so we could see better!" All the bedrooms in this house were upstairs. One night after Lucy had been quite

---

* Early in the war, the Japanese Navy placed nine submarines along the U. S. west coast from Canada to Mexico. Many ships along the coast were attacked and some were sunk within sight of land. On June 21, 1942, one submarine shot at Fort Stevens, at the mouth of the Columbia River causing no serious damage. On September 9, and again on September 29, 1942, the Japanese aircraft-carrying submarine *I-25* launched its bomber aircraft which then flew over the Siskiyou National Forest, dropped bombs and started a forest fire. Later in the war, the Japanese bombed Oregon and killed children who were on a picnic. The object of most of these raids was to stop the American timber industry by burning the forests to the ground. For complete details and pictures refer to bibliography for *Silent Siege III: Japanese Attacks on North America in World War II.*

difficult, George said; "Lucy you've been bad so you have to stay down here, the rest of us are going up to bed." Lucy said; "The shadows of the night were so frightening that she was afraid to go to bed." Next morning, George found her curled up in the big chair, freezing cold. They never left her down-stairs at bed-time again. Lucy does recall sleeping down there for several nights once when everyone except she and George had the measles. They figured sleeping up-stairs was a sure way to get the measles so they camped out in the living room for several nights.

Mr. Hunter wanted to make sure that his grand-children had a fun time. As soon as they moved into the house in Old Waldport he built them a set of teeter totters especially for little Georgie and his friends, but the big kids liked them too. What Joyce and Lucy liked best of all were the stilts that he made them, they had large toe straps to help keep their feet on. Helen said; "I'd be looking out the kitchen window while I was doing dishes, there'd be the girls walking right over the teeter totters on those stilts. That was so funny!" Every-one in town thought those were great. The policeman was so impressed with those stilts; he made a pair for himself. His name was Mr. Rhodes and his wife was Joyce and Lucy's teacher. Lucy said:

> **Grandpa built us kids a big swing there in Old Waldport. All the neighborhood kids thought that was the best swing they'd ever seen. It was a huge rope hanging from a great big tree. The seat was a gunny sack filled with straw tied onto the end of the rope. The big kids would stand up in the windows of that old vacant house so they could get a good start for their turn on the swing.**

Their grandfather built them a "Giant Stride" at their home in Gellatly Canyon. That was a large wheel that spun around on top of a pole cemented into the ground. There were seven long ropes attached to it. That was so Bill and Doris' five children, as well as Joyce and Lucy, could all play on it at the same time. They would grab hold of their rope and run around in a circle as fast as they could go. Soon they would begin to take bigger and bigger steps thus, "Giant Stride". When they got to going fast, they would be flying through the air hanging onto their rope. It was great fun!

Joyce seemed excited about going to school there. She met some girls her age and they became friends. Lucy being younger, was petrified by the thought of going to such a big school. In their little country school, there were only fourteen children in the entire school, often the older children helped the younger ones with their work. In the school in Waldport, there were thirty children in a room. At least her sister was over there on the other side of the room since the fifth and sixth grades shared a room. The schools in Waldport had six weeks or mid-term-tests, Joyce and Lucy hadn't had tests in their earlier school. Lucy heard the other children talking about how hard those tests were. They were studying the Civil War, its battles and generals and about Betsy Ross, the designer of the first American Flag, they also had to

know the states and their capitals. The other children at school told her that if she failed the test she would flunk, then she'd have to take fifth grade over. Lucy was mortified. She studied hard every night. After the first mid-term test was over, Lucy found out she'd gotten one-hundred on it. Mrs. Rhodes, her teacher stood up in front of the class and said; "This little girl from the country did better than anyone else in class." Lucy was certainly relieved to find out she wouldn't flunk fifth grade after all.

There were many interesting things to do in Waldport. Joyce and Lucy could walk to the tiny library for books to read, or walk over to visit their new friends. Sometimes they would run to the store for their mother or just go window shopping at Bogg's, where they had such interesting Christmas cards. Joyce liked the birthday parties best. Some of their friends included Mary Gourley, their closest neighbor, Betty Lou Hall, who's parents owned the grocery store, and Laura Lee Gilbert, who's parents owned the drugstore.

There were always new movies showing at the old movie house there in Old Waldport. Helen would cook a big Sunday dinner, then the family would go together to the movies. They were John Wayne fans. In those days it only cost a dime. Those were the days when you paid your money outside and inside they would sell you popcorn to eat while you watched the movies and they always had the news reels. These showed actual movies of planes dropping bombs and other battle scenes, that could be frightening to a young child. Then there were a few cartoons before the movie started.

In the summertime, the children went to Vacation Bible School. The girls took their piano lessons from Mrs. Leper and in the summertime they went twice weekly. For entertainment, there was always the skating rink, they loved to go there. At their Gellatly Canyon home they skated in the basement, it was more fun skating in the big skating rink with their family and friends. Georgie was only four but he remembers that sometimes his big sisters would take him along when they went skating so he could watch. Sometimes they would take him down to play on the beach near the new bridge. He loved to make sand castles and watch the water fill them up. Joyce and Lucy learned to swim in the small stream left there in the bay when the tide was out.

That spring, the school children had their class photos taken near the new bridge spanning Alsea Bay. (The dedication ceremony occurred June of 1935.) The completion of the new bridge at Waldport was an important step in the war effort as it was one of several large concrete bridges built along the Pacific Coast Highway. This allowed the flow of men and materials necessary for the war effort.

> It took many people to get through that war, it wasn't won only by the people fighting in it. All the civilians at home were doing their part too, lots of us were buying government bonds to help pay for the war. Mama and I bought some "Victory Bonds" every month. The smallest

ones cost eighteen-seventy-five and in ten years they would be worth twenty-five dollars; we could buy larger denominations too. I remember Mama and Grandpa raised 'Victory Gardens.' The government was promoting those, wanting everyone to raise a garden to help feed the country during the war. Mama and Grandpa had their first Victory Garden there in Gellatly Canyon. At the coast it was kinda hard to get anything to grow there. They didn't have good ground for their garden.

I remember we had ourselves quite a time over that sugar rationing. Sugar was pretty hard to get and Grandpa loved sugar. At one point, we got to fussing a bit over that. We were allowed five pounds per-person per-month. In the summer Mama canned alot so that didn't leave much for eating. We thought Grandpa was eating too much sugar, so everyone got their own quart jar and put their name on it. Grandpa was out in seven days, so was Mama. I was doing pretty good with mine though. So, Grandpa and Mama were borrowing sugar from me before the end of the month.

We played a lot of tricks on Grandpa. When we had pie, I'd ask Grandpa how much pie he wanted. He'd say; 'Oh, just a sliver.' I'd cut him a tiny piece of pie. He'd say; 'Oh! More than that.' Then I'd say; 'Dad you said, you only wanted a sliver.'

That summer, Grandpa went to visit Bill's family who had moved over to Cottage Grove where he was working in a sawmill. Grandpa was supposed to ride the bus back to Newport and be in about six o'clock. It was seventeen miles from Newport to Waldport so we went to pick him up Saturday night; but Grandpa wasn't there. After we'd asked around and nobody had seen him, we figured he'd just stayed a day longer so we went on home. The next day, about ten o'clock that morning he comes walkin' in. He'd walked that whole distance. He told us he'd walk awhile, then set along the road. We asked him why he wasn't at the bus stop. He said; "I'd stepped across the street to get 'a spot of tea and a slice of pie.' His Scottish background caused that. He was born in 1861 and he was 82 years old. Grandpa's 'spot of tea' cost him a seventeen mile walk.

Mama fed him his breakfast and put him to bed. Before that happened, every night when I'd come home late from work, they had to wait supper. He didn't like that, wouldn't talk to me much. Only thing he'd say was. 'Your late, ain't ya'? So when he walked in that morning, I said; 'Dad, you're late, ain't ya'?

Later, when Georgie and his friends grew up they use to tease Grandpa. When he went out to feed the animals, they'd go along ahead of him and turn the sprinklers on him. Boy! Would he cuss.

(TOP TO BOTTOM) **Settings tongs at Eckman's Slough. An early Cletrac in background. Loading a 1-log-load (Spruce) onto a 1941 Ford truck. Tongs being removed from log.**

> He'd have killed them if he could have caught them. I never liked to use dirty words; but I don't mind hearing them once in a while. Old Grandpa, he could sure cuss. I remember Mama telling the girls when they were little; 'Now, when Grandpa starts cussing, you girls are supposed to cover your ears and run away.'

Rex called George one morning while they were still at the coast. He wanted him to drive up to the Railroad Grade to start a gas Cletrac 76 they were having trouble with, George understood these well. Helen said:

> The men hadn't been able to start that cat for several days. Since the girls were in school Georgie and I went along for the trip. When we got to the Railroad Grade we walked on over to the cat. After looking at it a while, George climbed onto the cat. Georgie, only four years old, started running away. One of the men asked; 'What is he running for'? I said; 'He doesn't like the noise of the cat.' That guy said; 'Well, that cat's been pretty quiet lately. Ha! Ha! Ha'! The next thing we knew the cat was running. Georgie stopped and turned around just in time to see the cat puffing out blue smoke. Rex and Georgie were right, they knew George could start the cat.

George and Helen moved their family back to Gellatly Canyon after a memorable year at Waldport. Rex expected George to supervise the logging crews near Alsea.

The Castle

The Castle can be viewed at 24690 Alsea Highway, Philomath, Oregon. The grounds are open to the public for a minimum charge. The house is not open.

CHAPTER 15

SHROYER'S NEW HOME

Saving money was the most important factor in attaining the financial stability the Shroyers enjoy today. Helen's economic course in high school provided her with the insight that prompted her to place saving money above all creature comforts. Helen's visionary position regarding the use of money; combined with George's ambition and belief in providing responsibly for his family, insulated them from the depression driven poverty most families experienced. Their cooperative efforts made it possible for George to run a flourishing business for many years providing work and businesses for their children and many more. This allowed their family to live a vibrant, gratifying life.

While we were living there in Gellatly Canyon, I got to thinking that it would be a lot easier if I'd just go to work for myself. Then, there in the spring of 1945, a 'For Sale Sign' went up at 'This Place'. I got to thinking that it wouldn't be a bad place to live; I went by it twice every day on my way to the coast. I got to thinking that maybe Mama would like it and it would be alot closer to my work. That was another thirty-minute drive for me each day when we lived over in the canyon. After the 'For Sale' sign went up in the Thomsen's yard, I stopped in there and ask him if he wanted to sell it. He said; 'Yes, he wanted to sell it.' He didn't show much interest though.

I ask Mama to come and look, but she didn't want 'to come here at all.' She had a nice little place out there in Gellatly Canyon, she had a good garden and a new orchard. Finally she agreed, provided I would buy the 'Old Place' back at the end of five years, I thought that was kind of silly. I figured I'd make enough money logging for myself so I could retire in five years. I agreed to buy the 'Old Place' back. That goes to show you just how crazy a guy can be, can't believe I thought I could retire in five years.

Mama came in with me to see Mr. Thomsen on a Sunday afternoon. I remember him setting; 'Right here in the kitchen.' He never asked us to sit down or anything. I thought it was kinda funny he wouldn't let us sit down. Finally, I decided he figured we couldn't afford to buy it. We didn't have very good clothes. So I told him; 'If my wife likes it, we'll give you cash for it.' Boy! He jumped up off there, pret'near knocked us down. Started right in showin' us everything. Mama liked the place, so we paid him cash and that just about used up all our money. He'd wanted twenty-one-thousand for this place, few could afford it, that was a lot of money back then.

When we went back home that afternoon, we were kinda laughing about it while we were telling the kids and Grandpa; 'About that guy not thinking we could afford the house"! We'd been saving our

money, so we didn't have nice clothes. Our old 1937 Chevrolet wasn't very good looking; so we'd borrowed a car from Ernie; 'To come in here and make a better impression.' That didn't seem to have impressed him; but when I told him we would pay him cash 'He came alive all right'! Grandpa Hunter was kinda solemn, he looked me right in the eye and said: 'Well, you don't look very flush to me'!

George and Helen moved into the Thomsen's house spring of 1945, the same spring President Roosevelt died. The Shroyers' have lived there for fifty-three years. From the road, the 'Home Place' can hardly be seen. Once you pull into the driveway your eyes fall on the most beautiful 'Castle-like-Home' one could ever dream of. The evergreen hedges that provide privacy also protect the family against road sounds. The home sits on a grass covered mound of earth within a dozen oaks. These planted in three's, are providing a canopy of shelter from the summer sun. They enfold the surrounding fifty feet of the entrance running downward toward the creek on the left. On the right the earth slopes downward toward a picturesque willow tree planted in 1940. This opens into an expanse now filled with a spectacular 'Old English Garden.'

Once inside, George tells me, they were ushered to a screened porch running along one side, above the creek. Inside, a gently curving stairway hugged the circular form of the structure. They were then escorted to the bedrooms upstairs. George and Helen have made significant changes in the home, adding bedrooms where the porch once was. Their ever expanding family filled these rooms with laughter and a vigorous love of life. This home has sheltered them through fifty-three years of happy-times and tragedies, befalling this family throughout their lives.

The man that designed and built the home, built it above flood level. The expansive fields across the creek were lower than the rise the house stood on; thus insuring the home would be protected from flooding. George was impressed by the creek's dam and generating system providing the home with its own electrical power. This, and the large expanses of land well above flood level far from the house where they could build a barn and shop, these qualities interested him most.

Mama liked all the sun that we had in here, she didn't like the long winters we'd had there in Gellatly Canyon. In the winter time, the sun didn't shine in there much after two o'clock. The snow stayed on longer there in the Canyon, once, it stayed there two or three weeks after it was gone in Philomath.

After about six or seven years, I came home one night and I told Mama: 'I bought the 'Old Place' back. She said: 'What did you want to do that for'? I said; 'You made me promise to buy it back. She said: 'I'm not going to move back there'! She had a fit and started bawling around. I started to laugh. I said; 'I just wanted to see what you'd do if I bought the 'Old Place' back.' Mama just about killed me over that!

Twenty-five years later they bought back their 'Old Place' in Gellatly Canyon. The state purchased the land, intending to widen the highway to the coast. Shroyers had their 'Old Place' moved. Their old home constructed wholly with Grandpa's superior craftsmanship, now sets on a hillside not far from their 'Home Place.' It now protects a younger family of Shroyers. These walls have witnessed the laughter and the tears of four generations of Shroyers.

Helen's love of flowers continues to play a major part in her life. She is the only surviving charter member of the Philomath Garden Club she helped organize in 1946.

A poem was given to her by her friends from the club:

To Helen

Her garden is her hobby;
The flowers blooming fair
Reward her for the hours
She spends working there.

Her knowledge born of doing
And experience of the years
Result in floral beauty
There are many that it cheers.

Our garden club is richer
With Helen on our role.
She makes her yard a haven
As she gives her home a soul.

Little did Helen know that her love of flowers, and her industrious nature would, find a place for realization in their new place. Here, her gardening would mature into an immaculate, imaginative, inter-planting of seasonal varieties creating intricately sculptured seasonal gardens. Her elaborate ten acre garden mixes formal rose gardens, "Old English Gardens,, and quiet pools with a tasteful interspersment of conifers from around the world. A collection of two-hundred or more planted over the last fifty years. She has secured perhaps the first, of certain species of conifers planted in the state of Oregon.

Her first formal rose garden edged by beautiful climbing varieties leads the understanding eye toward one end of an oval lily pond, some one-hundred and

One of many spectacular vistas of Helen's garden. George and Helen's grand daughter, Kim Patterson Hill married James Holt in 1994 in the garden. George and Helen are seated together on the bench on left.

(TOP) **A young Helen in her new garden.** (LOWER, LEFT) **George preferred not to work in the garden.** (RIGHT) **Mr. Hunter, Helen's father, doing barn chores.**

fifty feet in length, filled with small fishes and oxygen producing reeds. This is transversed by a small formal bridge leading to the other side, allowing the observer to wander along its bank uninhibited by the Old English plantings surrounding the area. There must be a minimum of fifty vantage points throughout the garden, each providing the observer with a perfectly sculptured view.

Through the property flows a small river, whose banks are reinforced with large granite boulders and rocks, placed there by George Newman and his crew. A couple of substantial bridges span this gently flowing summer stream. So unlike the raging torrent it becomes when spring rains fill it till it seems, it would break out over the lower five or six acres and race rapidly over the remaining quarter mile of fields to rejoin the main river.

A climbing rose found in a private place toward the back of the property cascades some twenty feet out of a large conifer displaying it's delicate white flowers against this dark green framework, a beauty of incomparable proportions. Near this point stands a gazebo with wicker chairs, from the gazebo, one can view the formal miniature rose garden; a small offering of golden varieties of conifers as well as the terraced rock gardens on the far side of the creek. Additionally, a pond fed from above and a bit of the green house that contains ferns, palms, orchids and other tropical climate plantings.

Helen's genius, protected and unencumbered by her husband George, has fully developed here along the banks of this small river. It ranks with 'Buchart Gardens' of Victoria, British Columbia in complexity and design. Only a man secure in his own self and possessed with a love for his wife that overshadows his own personal needs would unselfishly, allow her genius to thrive and develop into this marvel of gardening. Only a genius could manage ten acres of innumerable species of plants with only a few persons working the grounds.

George is to be held in great esteem. Only a great man would provide the atmosphere that would allow his wife's genius to fully develop. He attained what all men should strive for, a deep love of his wife and family. His genius lay in his understanding that she would, as his help mate in life, function best if allowed to succeed to her fullest capacity. His genius included accepting his workers and respecting their faults along with their strengths. He, certainly as you will see in the remaining chapters, attains some remarkable accomplishments.

CHAPTER 16

STARTING-OUT AS SHROYER LOGGING

George wanted to start his own logging business and Helen encouraged him to do so. Helen and her father started a large garden in the sand loam on the other side of the creek. Helen thought they could sell the cream from the six cows that she and Lucy were milking every day. Mr. Hunter hadn't been well the past few years, but moving was an exciting time and he now had a 'new lease on life.' Helen bought a dozen White-Leghorn chickens for him to raise; she would be able to sell some of their eggs. The girls helped her sell the tuberous begonias she grew; all this helped with the household expenses. They would be struggling to get ahead financially; since purchasing their new place and the surrounding property had taken most of their savings. George's new logging business would require him to purchase several pieces of equipment and these added expenses would make it difficult for a while.

> **It was a lot easier once I started my own business. At the end of the day, I could say; 'Lord, I've done my best today—now it's up to you.' It felt good to be on my own, to live with my own mistakes, and try to do better next time. When I was responsible to Rex Clemens I was always real worried. I had a lot of headaches when I was working for him. Once I started my own business and I was responsible only to myself, I never had another headache.**
>
> **We started logging by Rex letting me buy some equipment from him and making monthly payments. That was just about the end of the war and you couldn't buy much then. First, we bought a second hand K-7 International truck, and an old D-8 Caterpillar. That fall we bought a Skagit-BU-150 to yard with during the winter months. Rex gave me a contract logging a piece of his timber, we were logging it off the Hayden place, three miles east of Alsea. At first we were cat logging, we only had three men to help me, Ernie drove the cat and Tom Coon set chokers for him while Jiggs Bennett, Bob Bennett's son, drove truck. We paid Ernie two-dollars an hour, and Tom and Jiggs, one-seventy-five an hour; this was a quarter-an-hour more than Rex was paying.**
>
> **In those days, we got eight dollars for logging and loading one-thousand board-feet. We got ninety cents for falling and bucking one-thousand board-feet, that was forty cents for falling and fifty cents for bucking. They were using cross-cut saws for falling and bucking, power saws were just starting to come out. We bought three Dimond-T's that next spring, there wasn't anything else available. They cost about nine-thousand-dollars apiece, now they are one-hundred-thousand dollars, but you get more too. The factories that made Fords, Chevrolets and Internationals hadn't finished retooling to**

(TOP) **A war-surplus (Army) International truck owned by Rex Clemens at Lew Holter's place, Alsea, about to haul a Douglas fir 1-log load. Alvah Hinton, driver, removed the side panels of the hood to aid cooling.** (LOWER) **George converts a steam donkey into a yarder by adding two GMC diesel engines. Used on Missouri Bend area.**

make vehicles for civilians. After that, we were taking out fifteen-to-twenty loads a day, a load would be about three-to-ten logs. We did that with just two men logging, besides myself. I would saw the limbs off the logs, measure the logs and put the size of them down on the log tickets. There was always plenty to do, they'd go home at night and I'd go ahead and log enough to load out in the morning. Anytime anyone wanted-off, I'd do whatever they had been doing.

Whenever George faced a difficult decision, he relied heavily on the enormous amount of experience he'd acquired throughout his working life. These past-twenty years had provided him with a broad base from which to work. He was able to start his logging operation with the bare minimum of equipment. George and Helen had used up their cash reserve buying their new home and the mechanic's shop for George. Reflecting on the early days, George realized that he could utilize the old method of loading logs from a rollway rather than buying a loading-donkey.

We couldn't afford a loader, so we just made what we called a rollway, we were loading all the logs with a cat. If the rollway was built right, you could roll a log on by hand with just a peavy. Lots of people today wouldn't know what a rollway was. We were using those rollways in 1928 when I got my first load of logs for Old McCoy. That was the time I loaded those logs with a peavy from a rollway built for loading with a team of horses. Ha! Ha! Ha! They sure had a good time over that, it was pret'near impossible.

One afternoon Georgie and I drove out to our logging site. Often I worked bringing in logs after my crew went home. It was about three-thirty in the afternoon when we met Rex coming in at the bottom of Alsea mountain. We both stopped our pickups and started talking about how things were going. As we were talking, I picked up a loose telephone wire that was laying there along the road side. They were stringing a telephone line into a farmer's house near there. Old Rex was talking away and I was just kinda listening, I figured I'd just fasten it to the back of Rex's bumper. He couldn't help seeing what I was doing but he just kept talking. I thought pretty soon he'd start giving me heck for that and make me take it off, you know. All of a sudden, he jumped into his pickup and was gone before I could unfasten it. Boy! I sure didn't mean for that to happen! Rex dragged that wire clear into Philomath. That night I heard from Old Man Langdon, he owned Philomath Telephone Co. He said; 'One of my men said he was up a pole and all of a sudden the wire took-off. He figured I needed to pay for that wire, I didn't argue with him about that. Rex'd given it back to me coiled up in smaller chunks scattered all over Mama's lawn. You could take the tires off Rex's pickup with him watching, he was always so interested in what he was talking about he never noticed what was going on around him.

Another time I had some fun with Rex, was when I was out in the woods building roads. He would drive out to look over timber and leave his pickup parked where I wanted to build road; then he'd go off and cruise timber. At first I got to working around him or moving his pickup, one day I got to pushin' dirt up against it. One day I did a real good job, covered his pickup with that dirt clear up over the tires. After that Rex got a lot smarter, he figured out where he ought to park his pickup. When I was working for McCoy and Clemens they put up with my pranks, they never fired me but Rex threatened to kill me pretty often.

I learned alot from talking to T. J. Starker. Rex and I often saw him when we were cruising timber. Starker taught me some about business dealings. He would always say; 'George you trust everybody too much. If you are in business, you can't trust everybody. After dealing with someone, if you learn that you can trust them, then—you better still watch them.' I owed him alot for his good business advice. Wished I'd listened to him more, he told me to keep my land after I cut down the timber, sure wish I'd kept it. Starker was always real serious, I never pulled any pranks on him.

The guys George hired that were returning from the service, were hard-working, dependable, and toughened, having faced the rigors of war.

Ermie Walter and Cotton, his brother, were the first returning veterans that we hired. We hired them when they first came home, that would have been right after the war, early in 1946.

They'd started falling and bucking in the early Thirties when they were working for Rex there in Newport and Siletz. They worked for him until they were drafted into the service. In those days the government classified the men as to their age, family situation and industry they were working in. Ermie was the first draftee out of Lincoln County to go into the war, Cotton was drafted later. When they'd been falling trees before the war, they were using large cross-cut saws known as 'hand-fiddles' or 'misery-whips.' Later, in the early Forties, the first power saws were available, some of the better power saws were Distons. I had one of those early gas-powered Diston saws, I'd asked Cotton to run it for me. Cotton said: 'No! We can fall more timber by hand than we can with that thing.'

Ermie and Cotton worked as fallers all their lives. They've fallen as many trees as anyone in this area. They would often fall over twenty-five-trees a day. One time at Yew Creek, Ermie and Cotton fell an Old

(TOP) **An early gasoline-powered Diston chain saw. Ermie on left; Robert "Cotton" Walter on right.**
(LOWER) **George Shroyer and Rex Clemens.**

George's first truck, a K-7 International carrying an old growth 1-log load – 1940's. Georgie at left, George, right. (INSET) **Georgie liked to be with George from an early age.**

Growth that was over four-hundred-years old. It was almost three-hundred-feet tall, and eight-feet-two-inches on the butt, this was the largest volume tree they ever cut. It was bucked into lengths containing two-hundred-thirty-eight log-feet with thirty-thousand-board-feet, enough lumber to build three houses.

Ermie said; 'The tallest tree I ever fell exceeded three-hundred feet. Its log-length was two hundred-seventy-seven feet and it contained twenty-two-thousand-board-feet. That Old Growth Douglas Fir was taken out of the Prairie Mountain area, soon after I came back from the war. Another tree I fell, was not as tall, but it had the largest butt-cut. Its inside bark measurement was ten-feet-eight inches on the butt, this was also an Old Growth Douglas Fir taken out of the Alsea area.'

Another veteran, David Rae, worked for me right after the war. He wanted a job driving truck. When he asked me for a job doing that, I told him I didn't need a truck driver but he could set chokers for me until I got another truck. He sure wanted to learn how to drive a truck. Pretty soon, I ordered a new truck then, I let him ride along with one of the truck drivers. It wasn't long before he was driving that old truck of ours. When I got a new truck, usually, a man already driving for me got to drive the new truck.

David Rae was doing pretty good until one day he'd loaded the thing up and he pulls off to the side of the road to get out of the way of the next truck needing to load. In those days, you had to keep the engine running to keep the vacuum brakes on. After parking that truck, he jumped out and ran back to the landing to get the ticket he'd forgotten. The engine died and the truck went over the hill. I'll tell you; he was scared plumb to death, that truck was wrecked real bad. I'd come along and was laughin', you know. I said: 'Well Dave, guess you are going to have to go back to settin' chokers.' In a month, I'd gotten three more trucks with trailers so I put him right on a new truck. I said: 'Bet you don't make that mistake again.'

Ermie Walter always said: 'There wouldn't be another guy in the world that would do what George did. Put a guy on a new truck after he'd just wrecked one. When something happened, George never looked around for a way to place the blame, or set a punishment. His main concern was always, well what do we do now to correct the situation. George was an extraordinary leader and role-model for his guys.'

I've always thought of Charley Kipper as my adopted nephew. He'd lived all his life in Missouri, until he joined the navy. He'd just gotten out of the navy when he moved here to be near Eva and Jess. Charley wanted to become a teacher and planned to attend Oregon Agricultu-

ral College and he needed a job to help him while he went to college so we hired him. Jess, his uncle, kept books for me. Jess was real dependable, so we figured Charley would be a good worker too. The second day on the job Charley was setting tongs and I was running the loader. We were loading Bob Bennett's truck. I didn't notice Charley reaching to adjust a cheese block on Bob's trailer when I put that log down. Logging is pretty dangerous, especially for a new guy. That crushed his arm real bad and he spent over a year having operations on that. Soon as he got out of the hospital and could drive again, we gave him the job of 'bull-buck.' *

The bull-buck determines the log's length and marks the cuts according to the requirements for that particular contract, he used an ax to mark these cuts for the buckers. He also did the wood's scale. This was a measure of the board feet in the bucked sections, with totals for each tree. The fallers and buckers are paid on this scale. We gave him an army jeep to drive back and forth to work. He worked for us full time in the summers. Charley was able to arrange to have all of his classes in the mornings, this gave him afternoons to work for us. Those veterans going to college were plenty serious about their jobs back then.

Fifty years after the veterans of World War II returned, the *Gazette-Times* in Corvallis printed a commemorative issue. This contained articles about veterans living in the area and one of these was about Charley Kipper. He'd never gone to high-school as he had being orphaned at an early age. He received straight A's during his first term at Oregon State College, the new name for the Agricultural College. Charley finished college in four years. He worked in this area and became well known, teaching both in Philomath and in Corvallis.

I hired George Newman right after the war was over. We always called him 'Punk.' He started working for me as a truck driver and soon, he was helping in the woods too. He started doing some of the high-climbing. Since my asthma was getting worse, I was real happy to have some help doing that. I remember Punk liked bananas and he always had a bunch of them in his truck. One day my cat broke down and I sent him in for parts. Punk locked up his truck so the guys couldn't get in there and take his dinner. We took those darn hinges off his truck, got in there and ate all his bananas up. When Punk got back and saw the door-off that, he knew what had happened.

Fred Day started working in Mama's gardens at first. We got to know him well and one day he asked me if he could work in the woods while he went to college. I said: 'You can set chokers for me if you want to.' His first day of settin' chokers, I noticed he was plenty clumsy out there. He was only wearing leather shoes with leather heels. He wasn't wearing any cork shoes. He broke his glasses the first ten minutes. It

---

* The boss of the fallers and buckers, also called bull bucker.

wasn't til later that I learned that his vision was 20\300 without glasses.

Fred Day recalls it this way: 'George took me into town and bought me the finest pair of cork shoes they made back then. The brand name was Curran, and what a difference they made in me. I put those things on and I was a new man. I could run like a gazelle, and when my foot landed, it would stay there. I got pretty agile out there when I could see and had a pair of boots on like that.

George said: "Fred worked hard after that and even though he moved to Texas later, he still writes, and visits us pretty often."

John Eagleson worked for me for over forty years. When he first asked me for a job, I told him if he could put the front unit on my D-8 Caterpillar, I would hire him to run it. He mostly ran the cat and yarder for me but could run any piece of machinery. He did some high climbing too. The machinery John ran, hardly ever broke down. He was easy on machines and always kept the gears and transmission greased. I always tried to hire people that were easy on equipment. If a guy could repair it then he usually took good care of it, you know.

John reminded me that he always worried he'd hurt somebody but he never got anyone hurt. They'd blow the whistle so he'd start yarding the logs in, the yarder operator couldn't see them, you know. Once in a while when they'd start up, the log would hit a stump and could come around and hit someone. John would worry that, if somebody got in the way, he couldn't stop it in time.

About this time we bought 'Superman,' the first portable loading rig around this country. We bought it from the army surplus place in Portland after the war. This Bay City truck was a heavy army truck, that had a crane and a shovel bolted to it's frame. I had Merlyn Couey drive it back from Portland. It created lots of interest coming down the highway. Besides being bright yellow, it had a large picture of 'Superman' painted on it. *Superman* comic books had just come out about that time and someone got the idea of puttin' that on there. Lots of people along the way stopped and looked at it; they were waving and whistling as he drove down the highway in 'Superman'.

We bought that in 1947, about two years after we'd started our own business. It was about the first of its kind in the country. I remember that Merlyn Couey had just graduated and we'd hired him to drive truck for us. We pulled that into our shop and Claude Baker helped me work on it. We figured we'd fix that thing up real good, make that into a loading rig. I was pretty good at fixing mechanical things and that crane with the shovel attached was a good piece of equipment. We changed that into a loader by adding a boom with a single tong on there. My uncle Tracy always wanted me to go to school in California. He thought I ought to become a mechanical engineer. Guess he was disappointed when I didn't do that.

First time I picked up a log with that shovel, it tipped part way over. We found out real quick that we'd better fasten it down. So, we built an A-frame over the center-pin. Now we could attach guy-lines to it and anchor it to near-by stumps. Later, we got to setting that thing up on two parallel logs, each about two-feet in diameter, that made it more stable while we were loading. We could pick-up pret'near anything without tipping over then, it sure earned its name all right. First time Old Rex saw 'Superman' he said: 'That thing won't last you two days'!

At first, we used 'Superman' at Five Rivers to keep the Camp Creek Sawmill supplied. We used it in the Alsea and Glenbrook areas after that. We used 'Superman' for ten-or-fifteen years – loaded lots of logs with it. Bud Carey and Cleo Duckworth ran 'Superman' for a long time. One man was needed to run the loader while another man set the tong.

One day when we were at Glenbrook, it was raining hard and Old Bud had been working out in the rain setting the tong. Bud and Cleo were always taking turns running 'Superman' and setting the tong. Setting the tong was pretty hard work, you know. You've got to climb up and down all day when you're setting the tong; to run 'Superman' you just run levers all day. They were waiting on a truck to come in so Bud climbed in to get dry. They started wrestling as loggers did in those days. Cleo wouldn't carry any cigarettes because he was always going to quit smoking. He was wanting a cigarette and Bud wouldn't give him one. So, they started wrestling harder than ever and Cleo shoved Bud hard against a lever. Bud says: 'I can't set tongs anymore, I'm hurt too bad, but I can run 'Superman.' Cleo just wouldn't believe that, so they went on wrestling for the right to run 'Superman' just as the fallers and buckers came by, Bud and Cleo came tumbling out of 'Superman's' cab. That night, Old Rex calls up and he says: 'Hey kid! What kind of an operation are you running a wrestling match or a logging operation.'?

Rex bought a contract on a section of standing timber near Missouri Bend. He'd bought it about three-and-a-half-years earlier. I had a contract with him to log off the last part of it that lay along the state highway on very steep ground. Rex worried that a log might get away from him on that, come down to the highway and kill someone. So, he sold me three-million-board-feet of standing timber for nine-dollars-a-thousand. His contract with Dr. Blower gave him four years, that was his last summer and fall to get the logs out.

We hired Joyce and Lucy that summer to do the flagging since we were falling the timber so close to the highway. We had just rigged up to haul those logs out when the weather became bad. It started freezing and thawing early that winter; so they'd closed the roads down to hauling. We had no way of getting those logs hauled out before the time would run out on the contract Rex had with Dr. Blower. Dr. Blower lived back in Ohio, he wasn't aware that any logs were still on his land.

One morning early, I got a phone call from Dent Perrin. He told me he'd heard Buryl Tom, his uncle, and Rex Clemens talking about buying that land at Missouri Bend to get back my logs that were on the ground. They knew the original contract was expiring, the highway had been closed to hauling and I couldn't get those logs out. They were sayin': 'Things are going too good for the kid. That's what they called me back then. They thought maybe they ought to teach me a lesson. Right away, I called up Ethel and asked if she had Dr. Blower's phone number there in Ohio. She was real helpful, gave me his phone number and address. I called him right up. He said; 'Another party from your area is interested in buying that land.' I offered him more money than they did and he said he would take it if I wired the money directly. My insurance agent, Jack Porter, helped me wire a check to him for nine-thousand-five-hundred and I had a telegram back in one-half-hour that said I owned it. Not one of those guys, ever said a thing to me about that. I knew them real well. I always figured someday Rex would say something about that, he never did. They probably decided that the kid had grown up.

Dent sure did me a good turn. Two-years earlier I'd helped him once. He was drumming logs up a steep canyon on the Snag Road. That road got its name from the old-growth snags left after a big fire that burned there about in eighteen forty. Dent had taken up all the smaller logs but his equipment wasn't large enough to handle the six-to-eight-foot diameter logs. He'd been talking about that, so one weekend I sent Ernie and Bruce over there with my two D-8's to drum up those logs. When Dent went to pay me I told him he didn't owe me anything. Perhaps you'll have a chance to do something for me someday. He sure did!

We cleared four-hundred acres of that section, bought some adjacent land to go with it that had the Alsea river running through it. We made it into a real nice cattle ranch for Mama.

Rex gave me an additional contract, so we needed to run another side. You couldn't hardly get any trucks at that time, because all the production was supporting the war effort. The army was selling used trucks to the guys that had been in the service, they had to wait six months before they could get the title. For anybody else to get a truck, they had to buy from one of those guys. Johnny Fields, our mechanic then, made a deal to buy five of those from a guy that had been in the service. So Rex bought one, I bought three, and Ben Ellis bought the other one. After we got those army surplus trucks, we hired three more truck drivers.

About a year later, the local newspaper came out with two huge headlines 'Three Prominent Loggers accused of Government Fraud.' Someone turned us in, I don't know who; never did find out. We thought that was pretty unfair since that was about the only way a guy could get a truck. All the production lines needed retooling to produce for the American public and that would take about a year. All our vehicles were pretty well used up, we didn't know about the six months waiting period so we started out using those army-surplus trucks right away. The war was over and we had all these guys returning from the service. We couldn't give them jobs if we didn't have enough equipment.

The day those headlines came out our attorney got a hold of us. He called Mama up and she drove down to Alsea where I was clearing our new cattle ranch and told me what had happened. She said they had already locked up Johnny and Ben. I didn't think anything was wrong, so I just went on home with Mama. After we got home, it wasn't long before federal agents showed up at our house. Mama answered the door and they told her they'd come out to take me to jail. They came right in here acting real mean. Mama said: 'George don't you pull any pranks, these guys mean business. Boy! Mama was right, they came right in the bedroom where I changed clothes, that's how serious they were. You'd have thought we'd robbed a bank or worse.'

When we got in the car, one was in the driver's seat and the other got in the middle and they put me on the outside. I started joking around, I said; 'It won't be hard for me to jump out. The guy driving got out of the car and came around to my side, so I had to sit in the middle, they'd taken me seriously. When we got into town I told them; 'I sure wish you guys would duck down or else let me, I don't want to be seen with you. It's a wonder they didn't shoot me.'

When we got into Corvallis, at the jail, this one guy had his revolver right along side of me and I could have gotten it. I thought about jerking it out and marching those guys right into jail. I knew the sheriff well, pretty soon I thought better; remembering Mama told me not to play around with those guys. I was in there about two hours before I could get hold of my attorney and he could get me out. Lots of business guys had bought those army surplus trucks. So, I figured I'd

write the names of all those I knew in the country that had bought those. I wrote their names on the wall of the cell. I wrote down poor Old Clemens and all the guys I knew of; I was writing all the time they'd spent in there, two-weeks, two-months, stuff like that. That wall read like the *Who is Who in Logging*.

I'd heard them say they were going after Rex Clemens next. Old Rex had just recently had a heart attack and I kinda wondered what they would do with him. Later, Ethel told me: 'Those guys came in there all ready to jerk Rex right out of bed and take him in. She called the doctor and lawyer and got them to say those guys couldn't take him to jail. Next morning, I went back to the jail and bailed Johnny out.

That hurt us all, some of the local guys took it serious and they were about ready to shoot us over that. One of my good friends never spoke to me again, guess they really believed what they read in the newspaper. I think it hurt Ben the most. He and Johnny went to church real regular, and I think their religious friends were 'too-quick-to-believe-the-worst.' They never seemed to really get over that, Ben died real young, so did Johnny, I think that incident caused them to die early.

We had to go to San Francisco two or three times to get that settled. Finally, the Judge got up and said if he locked us up, he'd have to lock up everyone in the United States who bought surplus, even for buying a tooth brush. He threw the case out of court.

Our new home had a nice shop, we did our repairs there and parked our trucks. We had lots of fun there. Johnny did most of the welding but Oscar Mock would help us sometimes. He had a welding shop there in Corvallis where Copeland Lumber is now. It was real greasy and full of soot in there, kinda like an old blacksmith's shop. Old Oscar wore wool clothing, since wool doesn't burn good. He looked real different in his old army khaki pants and shirt, and his wool scull cap that covered all of his head. He was real good at his trade so Rex always had him do his welding. When we started our own business we'd ask him out to the shop to do any welding that was critical since he was so good at it. Oscar built some of the first double axle trailers in the country. He was a great "Ford" man. He always wanted to use Ford parts. We'd asked him to come up to my shop one day to weld the case on a cast iron transmission for the D-8 Caterpillar.

We had him come over to the house to eat lunch with us and we talked about the fire he'd had at his shop. It had killed his parrot. He said that he'd gone to Portland to get a new parrot but he couldn't understand the new one. He thought it was talking Norwegian. I said: 'I've

got a Norwegian working for me. I'll send him in one of these days and you can have him listen to that parrot.'

So, I told Ernie to listen to what that parrot said, but not to tell Old Oscar what it was, I wanted to have some fun with him. Next time I saw Oscar, he said; 'What did your man tell you about that parrot of mine, could he understand him'? I said: 'Yes he could, but are you sure you really want to know?' I'd headed over towards the door—I was getting ready to leave. He was sure he wanted to know. So, I told him that parrot said: 'If this Old-Son-of-a-Gun would keep this place clean I wouldn't mind livin' here.' Oscar picked up his old Stilson Wrench that he always had on his lathe and chased me clear round the block with that.

We had lots of fun in my shop too. In winter when the logging was shut down I let the truckers work on their trucks, they could earn a check that way. One time Slim Davis was cleaning up his truck, getting ready to paint it. I was cleaning the stove pipe and Harold Fields yelled: 'Hey Slim, look at this'! I remember Slim looked up just as I shot inside that pipe with the air hose. He looked just like a darky, with only the white-of-his-eyes showin'. I said; 'Slim you better go to the house and see Mama, looks like you need to wash your face' It's a wonder he didn't kill me.

Slim and several of my truckers lost loads just as they came around a steep hill. We were hauling out of the Reed Road near the foot of Alsea Mountain. It was a tough, hard-haul with the equipment we had in those days. Slim was a good natured guy so I decided to play a trick on him. I made me a big sign that said; THIS DUMP IS CLOSED. Slim and the other truckers got the idea after that; we all had a good laugh, especially me.

CHAPTER 17

LOGGING, A DANGEROUS BUSINESS

Falling a tree of considerable size was a difficult task. In the first part of the century, the trees very large and the tools were primitive. There was usually a team of three fallers who would determine where to fall the tree, generally falling it in the direction it was leaning. Next, they would determine where to make the cuts. The accepted position for the cuts was above the portion of the trunk that spread out with the root mass. In large trees, this cut could be waist to shoulder high on a man, depending somewhat on the steepness of the surrounding area. On the front of the tree, a forty-five degree wedge was cut to allow the tree to fall in that direction. Chopping downward, they cut out a wedge aligning the horizontal part of the cut with the main cut. In the early days they weren't concerned with wasting a portion of the butt-cut. Today this wedge is called an undercut and it is sawed below the horizontal cut.

When the trees were exceptionally large the man on the lower side of the tree usually stood on a spring board. The spring-board was a rough two-by-six, four-and-one-half-feet long. The men chopped a narrow slit into the tree trunk at the appropriate position, placing a spring board in this cut. The man cutting from the lower side worked while standing on this platform. The man on the upper side of the tree, stood on the ground.

The main cut was made with an eight to ten-foot, two-handled saw known as a "Misery-Whip". This cut was a large horizontal cut made from the back of the tree, about half-way through the tree. As they were doing this, a third man drove steel falling wedges into the sawcut with a sledge hammer. Placement of the wedge was just behind, this kept the weight of the tree from binding on their long narrow saw. Driving in these falling wedges also caused the tree to lean in the direction the fallers wanted it to fall. These trees were so large that men dare not let them start leaning the wrong way. Today men use power saws and the trees are so small they never use springboards.

Their falling wedges are shorter and wider today because the trees are smaller. Sometimes wedges are not used at all. George said; "In the early days, a man expected to buy only one glut, or wedge, in his life-time. Later as pioneers moved west, Hickory was no longer available to them, they then began making a skeleton-glut out of a combination of iron and wood. The wedge's first six inches was made of iron, and attached to a skeleton that held a block of wood with a piece of thick iron providing protection at the opposite end. Therefore, the men in the west would use blocks of other hardwoods that were not as strong as hickory.

> **Dad told a tale about the early days in Kentucky. He said that one day my great uncle was out cutting wood for his family, when some Indians came out of the woods to capture him. There were about six of**

The fallers are standing on springboards using a "hand-fiddle" to fall this old growth Douglas fir. Possibly as late as 1930.

—Benton County Historical Society

them and they already had his family. He explained to them that he would come along willingly, if only, they would assist him in splitting this log he was working on. He had them put their fingers in the crack that had formed, then he hit the splitting-glut with all his might, causing it to pop out of the tree. The tree closed-up with their fingers still in it, then he took an ax and killed them. The Indians hadn't ever seen a splitting-glut before, so they didn't know that would happen.

The fallers and buckers work ahead of the logging crew, with the fallers working ahead of everyone. They normally would work about one-thousand-feet-or-more ahead of the buckers. Generally there were two buckers behind a crew of fallers. The buckers commonly worked separately, but near enough to each other that one could get help in case of an accident.

Old Man Hawkins, and George Huffman worked together pretty often. I remember seeing Old Man Hawkins, he'd go in there and jump up and down on those logs so he'd know where to buck. He was real good and would make his quota every day.

When the log is supported on both ends, sometimes, the bucker would start by cutting into the log from above, they would cut downward about eighteen-inches-or-so, or, until they felt the saw-blade starting to bind. Then they would remove the saw, and place their bucking-wedge in the sawcut, the cut was finished by sawing upward, from below. Pinching can cause splintering, and this would ruin the log, so buckers had to be real careful, it takes lots of experience to be a good bucker. I remember sometimes, when George Huffman was bucking, he'd get his saw pinched. When that happened, he'd have to chop his saw out with an ax, that could take all day. Bucking seemed like a thankless job; I'd go crazy doing that. Some guys kinda liked being out there by themselves all day long.

Accidents were common to bucking logs in the old days. Rex's crew was working a couple of miles above Dickinson's Camp on the Railroad-Grade. When the fallers got through in those steep canyons the logs would be lying massed on the ground. This made it dangerous for the buckers who cut the logs into the desired lengths. One time Rex's crew was going to move into a new landing so George and a couple of the men moved the loading donkey to its new position over the weekend. Fred and George chose a standing tree. George topped it that weekend so the crew could get the pass-block and top-guy-lines up as soon as possible.

Early Monday morning, I'd just hung the pass-block in a one-hundred-foot spar-tree and was riding the pass-line down. I heard this hollering as I was coming down, it looked like something had happened to one of the buckers. I dropped my climbing gear soon as I hit the ground, and took-off running toward the place I thought the

Setting chokers. Dave Main (LEFT), Sam St. Clair (RIGHT). Sam holds a "talkie-tooter" signal device in his left hand which came into being with the use of yarders after the era of the "whistle punk."

hollering was coming from. A bucker, Ray Pond, was buried face-down, in the dry moss and chips, under a large forty-two-foot log.

Some of the guys had seen this happen and had started trying to dig him out with their hands; I could see that would never work. Rex Clemens had started running back-and-forth, and up-and-down the road yelling 'I've got to get help'! That's what Rex always did; he couldn't figure-out what-to-do – he just knew somebody needed to do something.

The crew hadn't finished tying down the loading-donkey, so I yelled to Ernie to get on that cat, run one track onto the donkey's sled to hold it down. Vip was to get that line spooling-out! By that time, six guys hearing the commotion and came running over to help. I yelled to three guys to help me get this line-out. We grabbed that line and ran it out three logs past the downed-one. If we'd fastened it to the wrong-log, the one next-to-the-one burying Ray, that log would have rolled against it, killing him when Vip tightened the line. Instead, we wrapped and shackled our line to a distant log. Some of the guys had figured out what we were trying to do, so they grabbed that line and hooked it under-the-log that was on Ray. I yelled: 'Vip! Take up the slack"! We all held our breath, while that-old-log lifted right-off him. Some of the guy's close-by pulled him out where he'd be safe.

That had taken less than five minutes. In some places, those logs were about ten feet thick above the ground. We'd had to climb some two-hundred feet over a solid mass of logs, pulling that one-inch-line along with us. I remember thinking that we hadn't made it in time; that he didn't have much of a chance to live. Ray was a pretty lucky guy. He may not have lived even then, if it hadn't been for Old Henry Lein. Old Henry had just gotten back from spending a few years in the Marines where they'd taught him about artificial respiration. He got in there right away and started that. None of us had ever seen anything like that. Old Ray was kind of a purple-blue when Henry started working on him. It wasn't but about fifteen minutes before his color returned and he was breathing pretty good. Rex got Ray into his truck and headed for the doctor's office with him. Rex sure cared alot about the guys that worked for him.

During the early 1900's, large-spooling-mechanisms called drums, were powered by steam, they were used to pull the logs into a landing. They called these steam-donkeys and the guy running the donkey was the donkey puncher. The first donkeys only had one line on them and horses were used for pulling the line back into the woods. It wasn't long before they started using two drums, this allowed them to attach a haul-back-line so they no longer needed the horses to pull the line back into the woods.

The woods boss would say: ‘Take that log’! The chocker-setter was responsible for putting the chocker on the log and communicating with the whistle-punk. The whistle-punk would pull on something called a jerk-line. That would cause the whistle on the steam donkey to blow loud and clear. That meant the choker was around a log and it was ready to go. Soon as the donkey puncher heard the whistle he'd release the brake on both the main-line and the haulback-line. This was the old method of signaling used during the early part of the Twentieth Century.

When that whistle blew, the other guys had better get out of there. Those "Old Steam Rigs" were real fast. They were so-fast that the main-line was just like an elastic band – there was a lot of give in those things. They'd take up the slack, then suddenly it would tighten up and that log would jump right out of there, it was heading for the landing. You gotta' be sure you're out of the way when that happens. You could just hear those logs whizzing past you.

Chokers are usually put on the front, not on the back of the log. The thing to do is run back—never run ahead. If you run ahead, you're in trouble. In those days there weren't any safety rules, it was "each man for himself". Those steam rigs were real powerful, they had plenty of wood for the burner and a fireman that split the wood. The trouble was getting water to them. To run steam engines up there in the woods they had to get water brought in to them somehow, that could be real difficult. Often times they had to build a flume to bring water in, it could be a half-mile-or-more long.

Carelessness killed too many men in those days. One big logger I know was run out of the U. S. because sometimes he'd kill several men a day. He went to Canada where his heirs are still in business.

In the early twenties, they started building gas powered engines to run the drums, they now began calling them yarders. The first yarder I bought after we started Shroyer Logging Company was a Skagit BU-150, it had a diesel engine. In those days, that was a pretty powerful rig, it had a one-hundred-fifty-horse-power engine on it, now yarders have four-or-five-hundred horse-power engine.

Rex did most of his logging with cats during the early Thirties and Forties. After we started our own business, we did most of our logging with a high-lead-line and yarder. We used that old Skagit-BU-150 yarder. The main-line was a one-and-a-quarter-inch cable, the haulback was a three-quarter-inch cable and both of these lines ran through blocks that were up in a rigged tree. These lines were spooling off the drums on the yarder. The far end of it ran through blocks fastened to two stumps at the far end of the logging show. Chokers hold the logs to the main-line while we're yarding them into the landing. We were yarding in sections called roads. First, we'd yard in the closest logs. Then we'd work backwards toward the tail-block that held the haul-back line. Once we are through logging this, a five-sixteenths-inch haywire is sent out to help change the main-line over to a new section.

Loggers live by this rule: 'Never get in the bite of the line.' The bite of the line is a curve that forms in a line that is not tight. When the drums start to roll to tighten up the lines these curves straighten out. If you're standing in one, your life is measured in seconds; that's something every logger learns pretty quick. I'll tell you.

There at Missouri Bend, Bob Seits was running the swing-loader. The line had looped up on the drum and he was over there prying it out with a crow bar. At the same time, his brother was running the yarder. When he hit the haulback, way up on the hill, this tightened the line and it came up in Bob's face. It hit him on the jaw killing him instantly. That was a terrible tragedy, the first one we'd had since we'd started logging for ourselves.

They had a rather crude method of communication while working in the woods in the early days, they called it a jerk-line. They had simple commands then because they only had one line attached to the yarder to pull the logs in with.

This was used to communicate between the choker setters and the donkey puncher. The man that ran the jerk-line carried a line over the top of his shoulder and he'd jerk down on it one, two or more times. This caused the whistle on the yarder to blow. In the Thirties, when we started logging with cats we didn't use any signal lines. We just used one or two choker setters to fasten the chokers to the logs behind the cats, then the cat would pull the log to the landing. In the Forties, we started doing yarder-logging, we had diesel powered yarders. Then we needed two choker setters and a 'whistle punk.' The head chocker setter was called the 'hooker' and he'd holler the signals to the whistle punk.

This whistle-line ran from the diesel yarder out to where the whistle punk stood on a stump above the work area, he held the whistle-line in his hand. It had a squeezing mechanism on the end of it, they called this a "Bug" and when squeezed, it made contact causing the whistle to blow. The whistles were combinations of long and short blasts.

George gave some examples:

First—The choker-setter tells the whistle punk to send the signal

"Bucking a large log." Robert "Cotton" Walter uses a 40-inch chain saw, which could buck about 90,000 board feet a day.

Second—The whistle punk sends in the signal

Third—The donkey puncher hears the whistle and responds

Either: He runs the haulback-line to take the main-line back out to the choker-setter
-- Or -- He runs the main-line back-or-ahead and slackens it-or-tightens it.

The whistles:

One. if you were stopped—meant, to go ahead

One. if you were running—meant, to stop

Two. meant, for the haulback drum to run

Two and two. meant to make it come back easy

Two and three. meant to hold it tight

Five. meant, you are a "so-and-so"

Seven. meant, a man injured.

Different combinations meant different things.

Their communication system was continually evolving. These codes changed as did logging practices striving to fully utilize available equipment and technology.

On weekends George built the landings, when he did this in the summer, he'd always pick out a landing area that provided naturally humid conditions. The landing is a large flattened area where the logs are stored before loading on trucks. Summer-time often brought fires in the woods when the humidity dropped. Every eight-to-ten-years they would have especially hot, dry summers. When humidity dropped into the Thirties they would close the woods completely. A fire watchman lived in a trailer at the logging site. In the Eighties, loggers started keeping a person there all the time to guard against any one trying to sabotage their operation.

> **In the summer and fall, when there wasn't enough humidity in the air, we did something we called 'hoot-owling'. We'd go to work about four o'clock in the morning, whenever it was daylight enough to see. Then, we'd work until around noon or whenever the humidity dropped below thirty-percent. Often times, we'd be working there when everybody else was already shut-down due to low humidity. The fire wardens would come 'barreling-up-there.' They'd measure the humidity in the air and shake their heads, then measure it again. They couldn't figure out why it was plenty humid where we were. In the summer I would always pick a "naturally damp area" to lay out my**

landings. After 1980, they started measuring the moisture content in wood, instead of air.

After the Fifties it was popular to burn the slash. To do this, they put a mixture of gasoline and diesel in a special can that had a wick on it. The diesel keeps it from exploding and the gasoline keeps the flame burning. We'd pour this out where we wanted the fire to start. First, we did this to an area-up near the top of the hill so it would burn itself out. Next, fires were set-off at the bottom of the hill. The heat rising gives you a natural draft, this brings the fire up the hill.

When Georgie was thirteen-years old, he was helping me to start the fires. The fire below was coming up fast and I decided we had better get out of the way, we had to go over a hill. Georgie ran out of air before we got to the top and I had to carry him over. I was forty-five-years old then, and I had asthma. If I'd been ten years older, my asthma would have been worse and we wouldn't have made it. I shouldn't have had him up there with me, that was too dangerous.

High-climbing was real dangerous and Rex always hated to ask anyone to do that, he'd wait for someone to volunteer. I first started-high-climbing, out-there behind the Rat's Nest. That's the time I scraped most-all the skin-off my hands, and Ethel got on to me for using those words 'she didn't know I knew.' Ha! Ha! That was the winter, I started volunteering to do the high climbing for Rex.

When I think of climbing trees, I always think of Old Doc Hamm. One Sunday, when we were working down at Dickinson's, I'd gotten up pretty early to get set-up for logging the next day. I'd just put in a new landing and I was wanting to get some logs ready for Monday morning. I needed a choker setter and Doc was real good at that. So, I drove over to see if he'd go up there with me. It was probably nine o'clock and I could hear some yelling going on. They had a bunch of little second growth trees out in front of their house. Lorene was out there looking up in those trees. She had a big stick with a long piece of rope on it. She was flailing away at something up in those trees. I looked up there to see what was goin' on. There was Old Doc, he was naked as a jay-bird. He was real hairy, he was beating his chest, and yelling 'I'm Tarzan! I'm Tarzan'! He'd swing over into another tree when she'd try to get at him to get him down.

I laughed and laughed 'till I could hardly stand it. She said that he came home drunk thinking he was Tarzan and she'd been trying to get him down all morning. I had to wait awhile before she could get him down, dressed and headed out. Later, Old Doc was a high-climber for the "Bob Skelton Logging Company." The way he could swing around up in those trees was something else. I've never climbed around like Tarzan, but I've climbed and rigged about two hundred trees in my day. Of the spar-trees I've rigged, I'd say the average height was one-to-one-hundred-sixty-feet, loading trees were about sixty-to-eighty-feet high. When you are looking for a spar-tree to rig, you'd want to pick a straight tree that didn't have many limbs. It should have at least a five-foot-diameter base and you'd top it where it has at least a two-foot-diameter top.

Before you go up a tree you put on your climbing-spurs. These have spikes that are three-or-four-inches long. You kick these into the tree to hold you on there. Then there's the climbing belt that goes around your belly. It has loops to hold your hammer and spikes. Off my climbing belt I had an ax or saw fastened to the ends of two, six foot ropes. I carried one off my left side and one off my right. Most guys fasten a line to their belt, they'd use this line to pull-up the line for the pass-block, I never did that. I always carried up a ball of string in my pocket. I didn't want anything hanging from my belt that could get caught on something and put me off balance, that could be real dangerous. Some guys that carried their pass-ropes up with them would be carrying equipment weighing as much as one-hundred pounds.

When climbing, they also used climbing-ropes, fastening them to their climbing belt. These go around the tree to support them as they move up the tree with their climbing spikes. This rope is critical while they are performing the hazardous work of cutting the branches off these magnificent trees. While working their way up the spar-tree, they would flip their climbing rope up a little higher, then climb to that position, advancing up the tree. The men were working their way-up the tree where it would be strong enough to hold the rigging, yet high enough to give them lift. This section would have at least a two-foot diameter, that's were they topped it.

On some of the old trees, the bark had broken-off all the way around the truck. Sometimes, one of those tree trunks would be missing six-to-twenty-feet of bark down near the bottom. That old bark could come loose from those trees and you'd find yourself falling with the bark like I did once when I was working alone. The bark on that tree fell about twenty-feet that time, that was a real dumb thing to do. It's a wonder 'my time on earth' wasn't up that time.

When you came to a limb, you'd saw or chop it off close to the trunk. When I was up near the top, I'd always leave a stub of about eight-to-ten-inches on one of those limbs. I'd loop my climbing-rope over that while I was working up there, that made me feel a lot safer. I'd tie the pass-line over that stub after I'd pulled it up, that kept me from loosing it, while I was threading it into the pass-block.

Usually, I blasted the top out rather than sawing it because sawing seemed more dangerous to me. First, you'd decide where you wanted to top it; then, you'd notch the tree to make a place to put the powder. You'd have the guys down below attach the powder-caps and fuses up to a line so you could pull them up. That was kinda dangerous. I'd always set two fuses on a tree as I never wanted to climb that tree again to set another fuse. I'd allow myself about fifteen minutes to come back down, and get away from the blast, this would give me extra time in case my climbing-rope would hang-up.

Sometimes guys got killed going back up a spar-tree. They would wait for the blast to go-off, then they'd decide that the fuse must have died out and they'd start climbing back up the spar-tree. It'd go off while they were climbing, that would pret'near always kill them when that happened.

One time, a guy named Doc Picket, when he was only about thirty-years old, was setting on a stump making up primers for blasting. Gail, my brother, was helping him, and I was about two-hundred-feet away blasting stumps. That was when we were all pretty young, we were up in Henry Plunkett Canyon. He had a little chin pipe he smoked, he always kept it in the left pocket of his shirt. Usually, he kept the little caps that we used for blasting in his right pocket. One day he got careless, he'd been smoking his pipe while he was taking a break, and he hadn't cleaned his pipe clear out. He accidentally put his pipe in his right pocket with those little caps and it wasn't more than a-minute-or-two after he'd started working again that those caps blew-up; that blew his chest clear out. That was the first man we'd seen killed in a logging accident, that was really awful!

After blasting, it takes most of the day to finish rigging a spar-tree. First, you've got to attach all those plates that hold the guy-lines to the tree. You've got six-to-eight top-guy-lines to attach up there near the top of the tree, they will hold the main-line and haul-back-line up. There are three-or-four belly-guy-lines that go on the tree, about two-thirds of the way-up after the blocks are hung on the tree. These guy-lines help support the tree during the logging operation.

The most difficult part, hanging the blocks, comes last; there are straps and shackles to put on before each block goes up. The first block placed on the spar-tree is the pass-block weighing about fifteen pounds. After you've pulled-up its strap and shackle, and placed the strap around the tree you pull the pass-block up and hang it in the shackle. After that, you pull up the haywire up and string it through the pass-block. Next, there is the haul-back-block to add, these weighed about one-hundred-plus pounds. Of the three blocks that you'd be putting up there, the largest block, the high-lead-block, weighed one-thousand-pounds-or-more. The strap and shackle for it weighed at least one-hundred-fifty pounds. To put this around the tree you'd walk around it carrying this heavy strap. The worst part, was trying to line up the hole in this heavy block with the hole in its shackle while placing a two-inch pin into them. You only had about one-eighth-inch leeway when you were aligning them for the pin. Rigging-up a spar-tree was quite a day's work I'd say!

The high-lead-block was very different in those days, than it is today. Those blocks had two oil reservoirs and the sheave section was about four-inches wide, this rotated with the pull-of-the-lines. This section was floating on the oil in the reservoirs, each of these held a gallon of oil, and in those days, the seals were made of felt so they didn't last very long. We had to fill these reservoirs every two weeks for the newer ones and as often as daily when the bushings wore out. Today we use roller bearings instead of bushings, and these have cup grease in them, this is thicker and lasts much longer than the oil did. Today, they make the seals out of neoprene, so they last longer too.

One time when we were working up on Klickitat Road it was snowing hard and it was real windy. The crew on the ground helping you is real important, you communicate with them through hand signals and your voice. We used hand signals to let the men know how long the strap needed to be, the length of the strap is critical, it has to be just the right length. If you are going around a two-foot tree you need a seven-foot strap. If they sent up the wrong strap you had to send it back down, then they would have to try sending you another. The weather was so bad the men below couldn't see me. Since they couldn't see my hand signals, sometimes when they were sending something up, they would send it only half-way-up. With the wind blowing and machinery running, it can be very hard to hear them. We should have gone home 'til the weather cleared but that didn't pay the bills, you know.

I never felt safe, never liked doing that; always promised myself I'd never go up another spar-tree; but I did. Only knew one guy who really liked it, that was Old Fred Gee.* He did lots of climbing in the early days and I remember he'd always look down at me, while he was sittin' up-there smoking. He'd yell down; 'Hey kid! It's real nice up here, why don't you come on up'? Sometimes he would stand straight up on top of that. That wasn't the way I felt about it, I wanted to crawl under the bark while I was climbing a tree; never really liked it up there. I always rigged a spar-tree real good while I was up there, 'cause I knew – if something had to be fixed, I'd be the one that would have to go up there and do that.

---

* For picture of Fred Gee on top of a spar tree turn to page ix. (Editor)

CHAPTER 18

SAWMILLS, MINES, AND DETROIT DAM

George and Helen's girls were both in high school and had many friends. George and Helen recalled how serious they became about each other while they were young. They knew it wouldn't be long before their daughters would be looking for partners themselves. Georgie, their little brother, was too young for that; but he was beginning to play jokes on his grandfather and the girls whenever he could. George was just beginning to get his feet-on-the-ground with the "Shroyer Logging Company." He and Helen bought land near Alsea and started their Alsea Ranch. George and Rex were considering investing in a small sawmill in Lobster Valley and they needed another partner, one that would run the mill.

George, Rex and Ike Hovey bought the sawmill in Lobster Valley at Camp Creek that spring. They paid forty-five-thousand for it. They bought it from Mr. Miller and Mr. Fish. Rex put in quite a bit of start-up money, Ike Hovey managed the mill, and George had the job of logging Five Rivers to supply the mill with logs. They shared the profits from the mill equally.

On the home front, Lucy, George and Helen's younger daughter, was the first to pick out a beau. After the war, Mrs. Jessie Patterson and her two sons moved to Philomath. Her oldest son, Jimmy Patterson, served in the Navy's C. B.'s. These construction battalions built airports, and docks for the war effort. It wasn't long before he noticed Lucy, a good-natured, pretty young girl from the high school. He'd seen her at "The Orange Owl" eating lunch with her sister; and she'd noticed he was always joking with his friends, rather like her father did. It wasn't long before this young veteran from the navy asked Lucy to go to the movies with him.

> **One Sunday, Jimmy came out to see Lucy. He'd been hauling logs for me. We started talking about the new sawmill we'd bought there in Lobster Valley. I told Jimmy that we needed a good driver to haul the lumber from our mill to Ben Ellis' plainer in Philomath. Jimmy said he'd sure like to do that, so he went to work for us doing that.**
>
> **We were going to power the mill by using the two GMC diesel engines called "Twin Jimmys." When we put the "Twin Jimmys" into the sawmill, I saved the brass whistle that had been on our old yarder. It sounded just like a train exhaust. We'd used those diesel engines in the yarder we'd built when we logged at Missouri Bend on the land I'd bought from Dr. Blower. Those engines put out about 386 horse-power. That mill could cut about thirty-to-forty-thousand-board-feet per day with those "Twin Jimmys" and a crew of nine men.**
>
> **We built a one-man lumber-loader. We built that where we stacked the lumber as it came off the green-chain. Jimmy would back the**

George bought a mining claim in the Bohemia mining district of the Calapooya Mountains in eastern Lane County.

truck he was driving between two posts, knocking down the timber that held the lumber. Then rollers on the back of the truck would move the lumber onto the truck. That next year, we worked that one-man-loader with a couple of hydraulic jacks. Jimmy would back under the rack the lumber was setting on, then he'd lower the load down onto the truck bed with those hydraulic jacks.

The Five River Road was pretty narrow, it was a one-lane with turn-outs. We tried to use a trailer but it didn't work, so Jimmy hauled it "bob-tailed."* Jimmy worked long hours, weekends and into the night to keep that lumber hauled out of there.

Mama and I invited her brother, Bill and his wife Doris, to come camping over the 4th-of-July. That was the summer of 1948. We took all our kids and their friends camping. We went up into the Bohemia Mountains about twenty-five-miles southeast of Cottage Grove to look over the old gold and silver mines in there. Rex Clemens owned a few of those. Kenneth Watkins had an engineering degree from OSU specializing in mines. He was a timber buyer for both, Rex and I, and had agreed to show us through the mines up there.

Mama and Doris kept busy cooking all the good meals. They set up a kitchen there in that old mining shack next to the Sunshine Mine. The 'Old Timers' had worked these mines in the early days. The mine they called "Lena" was about three-thousand-five-hundred-feet down. You climbed fifty-section ladders to get down there. We walked a long-way back in there. We used little carbide lights, fastened to our caps, to see. There was a railroad down there with a track and cars that worked on air motors. We went down in the "Musik Mine" too. There were several-thousand-feet of tunnels under it. I never felt safe down there.

Bill's oldest daughter Maxine, brought her boy friend Gene Byler

* Six-wheeled truck without a trailer for hauling logs, was effective on narrow and steep roads with sharp turns. (Editor)

along. The kids and their friends had a good time playing games and singing. Gene played songs on his guitar, he could play real good. Jimmy could tell real good jokes, kept us laughin' alot. I got up early one morning, I remember rigging up a bucket of water over a doorway when no one was lookin'. One of the boyfriends got a cold shower with that. Boy! That was a real surprise.

Rex had seventeen improved mining claims he wanted to sell. He thought he might get in trouble with the government again, if he took the timber-off those. Kinda like the trouble we all got in when we bought those army surplus trucks. The old mining law of 1880 gave you the title to any claims you improved. If you did that, it was legal to take-off the timber; but there was opposition growing to that in Washington, D. C. Rex didn't want me to buy those claims, he thought I'd get into trouble over that.

Kenneth told me it would give me some real good timber and he'd been helping Rex buy timber and mines for several years. I hired Kenneth to buy those claims from Rex and I paid him thirty-thousand for seventeen improved claims. Kenneth helped me buy about a-dozen-more that needed to be improved. He said he'd improve those other twelve mines for me; but he didn't do that before they changed those laws.

I paid a mining concern in Portland to tell me how much gold, silver and copper there was in those mines. They told me it was there, but I'd spend as much as it was worth getting it out. I never did any actual mining, just took the timber-off, took that off about half-of-those by contracting an outfit to log that for me. During Eisenhower's first administration, they changed the law, and they made it so you couldn't take the timber-off mining claims anymore, so I sold my mining claims not long after that.

That was the first-quarter-of-a-million I'd made-off investments, that was easy, we didn't do any real work, just paperwork. I needed to reinvest my profits from that, so Mama and I started looking at some property and timber both in Idaho and Canada.

We were still running that mill there in Lobster Valley but Mama was after me to take the family to California for Christmas. I'd just put in a bid on a big government-contract for clearing a dam-site up near Detroit, Oregon. She knew if I got that job, we wouldn't have any vacations for a couple of years.

Mama was always wanting to go on a trip, she loved to travel. We'd go to Alhambra, California to see Uncle Tracy and my cousins Hazel and Ester. Hazel and Lawrence had a big family and Margie, their daughter, had a boyfriend who rode as a jockey at the Santa Anita

Race Track. Mama liked to go down to Santa Anita and bet on the horse races. I told her we could go for Christmas.

Lucy wanted to know if Jimmy could come along, kinda go along with the family. It looked like Lucy and Jimmy were getting serious; so, Mama and I said he could come along but we didn't have enough room in our car. Jimmy talked a friend of his into driving with him, down to California; we took-off with two car loads for our vacation. Jimmy and Richard Roberts drove Jimmy's car, and we took our Cadillac. That was the first time Joyce met Richard; we called him "Rich." She sure seemed to like that guy, got him to throw his cigarettes away before we headed back to Oregon. Those kids sure had a good time. Jimmy was lots of fun to go places with cause he could tell the best jokes. Once we got to California we were driving along the highway gawking at this and that, taking our time—just like we did in Oregon. The highway patrol stopped us and said; 'Get those cars going, this isn't Oregon, you know.'

Georgie sure hated to travel; but once we got there, he sure had fun with Carolyn his ten-year-old second-cousin. On New Year's Day we went to the Rose Parade there in Pasadena. They were selling these cheap periscopes you could use to see over the crowd. I sure liked those roasted peanuts they sold there. Bought me a bag of those, then I went around and dropped peanuts in those viewers. Boy! That sure surprised them; I always made it look like Jimmy did it. Carolyn was wearing a coat with a long belt; I took-off her belt and tied her to a tree, everyone got a big laugh out of that. Mama always loved these trips to California.

We had a real good time that Christmas. Those were the days; life seemed pretty good back then.

After that, Jimmy kept delivering the lumber for the Lobster Valley Mill and kept his eye on it for me. Rex and Ike wanted to sell their shares in our sawmill in 1949 and we were able to buy them out with some money we'd made on the Bohemia Mines.

We bought a brand-new, 1949, K-12-International-Truck, early spring of 1950. Jimmy was driving that for me, we got that just before he married Lucy. It was a real nice looking truck. We decided that truck needed a whistle on it; I'd saved the brass whistle-off the old yarder so we put it on that lumber truck. That sounded just like a train exhaust. Jimmy and I would be at the foot of the Alsea Mountain coming in with a load of lumber, and we would blow that thing. The horses were out in the field, and those darn horses would jump right over the fence into another guy's pasture; scared those horses to death. The cows would try to run through the fence; but those horses

Clearing land about 1 mile above Detroit Dam. Eugene "Shorty" Fenner cat driver.

would just jump over it. We had lots of fun with that 'Old Brass Whistle.'

Lucy's oldest son, Randy Patterson, now has it on the flatbed truck he drives for his brother Jeff. He goes up and down Interstate #5 with that "Old Brass Whistle" blowing.

When we came back from our trip, the government let me know that I'd gotten the bid on clearing the entire north side, and the southwest end for the Detroit Dam. That was the first time I'd bid on a big government contract, they'd asked for bids for logging, and clearing the land. We cleared the steepest ground at the Detroit Dam site, and someone else cleared the upper southeast side. The timber we logged

was both, 'Old-Growth' of five-or-six-feet on the butt, and 'second-growth' that was only two-to-three-feet on the butt. To fall, buck, log and put it in piles of different lengths; they paid twenty-seven dollars for one-thousand-board-feet. That was the biggest operation we ever ran. We had twelve cats and two-hundred-fifty men working there one time.

The cook house at Detroit Dam was run real good. They fed at least fifty men pretty much all the time. They ate there regularly during the week. When the cook went home over the weekend the guys that stayed there would take turns doing the cooking. There would be about twenty guys there on weekends keeping the brush-fires going. One weekend, my brother Gail was in the kitchen taking his turn at cooking. The rule was; "If anybody complained about the food; we'd make them cook. Gail had his apron on, and he'd cooked a big pot of beans that were especially salty. Jimmy grabbed a big dish of them, soon as he tasted them he grabbed his glass of water, drank the whole thing right down without stopping. When he came up for air, he said; 'This sure is salty'! Gail started to take his apron off as Jimmy said; 'That's just the way I like 'em.'

Gail bought a couple of imitation 'Polish Hot Dogs.' He'd bring those things in, and put them on a plate right out there on one of the tables at dinner time. Some of those guys would see those right away, when they came in to eat and they sure looked good. They would make a dive for them, then they'd find out they were just rubber.

Colonel Beck represented the government on this contract; I got along with him real good. He had a son he was having quite alot of trouble with, and he wanted us to give him a job. I'd let on like I was going to; but I never did. His boy didn't like to work hard. Sometimes, Colonel Beck would come at meal times, when he did; he always ate with us in the cook house.

One time Oliver, my Dad, went up with me to Detroit Dam. He was real law abiding, you know; and me, I'm kinda the other way. When we were heading home, we went through a thirty-mile-an-hour speed-zone, where people usually went forty-five-miles-an-hour. We were traveling behind a string of about twenty cars, that were going pretty slow and I couldn't imagine why they weren't going any faster. No one was coming, so I passed the whole darn bunch. When I got right in front of them, there were two cop cars; they pulled me right over and gave me a ticket. Dad said; "They should charge you a-hundred-dollars for that.

I'd heard the judge up-there was real mean. He had no use for anyone caught speeding. When I got there, a dozen people were waiting to see

him. I'd heard he had some timber for sale; so, when his secretary wanted to know what I was doing there. I told her I was interested in that timber he had for sale, he told her to send me right in. They just left the rest of those guy's waiting; they just had to set there. I talked to him about his timber, I acted real interested in that timber of his. I told him I would have to go out to look at it first. Then I said; "By the way, I have a ticket here, what are we going to do about that?" He said; ‘I'll fix that’! He just tears it right-up and throws it in the waste-paper basket. I got out of that pretty good. Funny thing, I sure lost interest in his timber after that.

While we were logging timber we had to burn-off the slash, we kept our brush-fires going day and night, we never let those fires go out. The cat skinners had to be pretty good with a cat, to get those branches and root-wads piled-up right. They had to push those into piles, without getting much dirt in with it, or they won't burn good. Once we had a big fire up there, and it got-away from us, and it crossed both, the river and the highway. It burned up lots of government timber on the other contractor's side of the hill, five-miles away. Everyone including the government guys fought that.

After we got the fire out, those guys started putting all the equipment away, I didn't think they ought to do that. I wanted them to leave the equipment out, in case the fire started-up again. After a big fire like that, it could start-up again just about anywhere, if the wind came up. According to government regulations, we had to put all that fire fighting equipment away, wrap-up those hoses, and lock them up tight. That very night, the east wind came-up again, and before they could get all the equipment out, and in place—the fire had spread-out over the entire area. It started burning something awful, there were half-a-dozen logging-outfits in there, fighting that.

The government paid everyone for fighting that fire; out of the money they were paying me for logging, they should have billed me instead. My insurance would only pay for my losses; so if they'd sent me a bill for that, my insurance would have paid it. My attorney said it would be a long, drawn-out fight, so we never tried to get it back. I lost all of that money; that fire cost me half-a-million-dollars.

After we finished logging that, they put the logs up for auction. Those logs were worth about fifty dollars for one-thousand-board-feet. The owners of the sawmills up there were asking me if I was going to bid on those logs, when they went up for auction. I said; "Sure—I logged them, so I'm bidding on them. I'd never thought much about it before; but since I'd told them I would bid on it, I put in a real low bid

of seven dollars for one-thousand-board-feet. The saw mills thought I would bid too high, and it scared them-off. I was the only one that put in a bid. So, Old Beck, he called me up and he said; 'You come on up here; you got the bid on those logs.' I told him: 'I never figured on getting them; but since I did, you guys are in trouble.' Beck said; 'How is that'? I said; 'You've got another bunch of logs coming up that you will need to take a bid on, those will be up for bid in another month. What are you going to say when they get the next bid for sixty-dollars-a-thousand-board-feet and it comes out that I only paid seven-dollars-a-thousand'? He didn't know, so I told him I knew how he could get out of that. I said; You just put a rider on my contract to include that. Well, you know, in fifteen minutes they had a rider on my contract and they gave me the whole thing for seven-dollars-a-thousand. That kept them out of trouble and made me alot of money. It kept me from losing money on clearing that. I would have lost about half-a-million if I hadn't bid on those logs, as it turned-out, I came-out about even. I'd bid on two-and-a-half-million-board-feet and the rider added another two-and-a-half-million-board-feet of timber.

It took two years to finish that contract; everything had passed inspection except thirty acres, down along the river, that wouldn't burn. It was Christmas time and the weather was rainy and bad. We had lots of large root wads we couldn't burn; Colonel Beck and his bunch knew the root-wads were there. You could bury them if you didn't get caught; but they wouldn't tell you that, you know. The crew dug this big hole, two-hundred-and-fifty-feet long, sixty-feet wide and fifteen-feet deep. I had twelve cats all shoveling those root-wads in. We were just gonna cover those up with dirt when, wouldn't you know it—here came three cars of inspectors. They saw what we were doing but they waved at me, so I waved back, and they went right on by. I couldn't imagine them doing that. Going right on by like that. Boy! I told the guy's we better hurry that up.

Those inspectors stayed up there a long time, they never showed up again 'til we got it all covered up. When Beck came back, he said; 'George, I never noticed this hump before. I said ' It's been here since I've been here. That's the way he passed it—as long as I said it was OK—that was good enough for them. That covered their liability. They knew that stuff wouldn't burn 'til next summer, and I had to finish. I was losing money. It makes all the difference in the world what kind of inspector you get. I've always said: 'If he likes you, he lets you get by with murder; if he doesn't like you—he murders you.'

Later, I got to reading that contract I'd signed. In the bidding pamphlet I first read, it said; 'Everything ten-feet long with a two-

inch diameter had to be cleared, and burned. In the contract I signed, they'd changed that to say; 'Anything that would float had to be burned,' this was in fine print. Those root-wads wouldn't have floated, sure wished I'd seen that fine print when I first signed that contract.

I sure gained alot of experience doing that. After the job was over, they kept after me to come and work for them again. They offered to pay all expenses, and give me a ten-percent profit on their clearing jobs. Also, they'd sell the logs to me at a good price. I was so sick and tired of government regulations by that time, I wouldn't do it. I told them it wasn't right; for just one guy to get to learn how to do that.

CHAPTER 19

FAMILY MILESTONES

George was a very handsome and proud "Father of the Brides." He was much like John Wayne in stature as well as in character. George, reaching the fullness of his manhood, looked from a distance by his mannerisms size and appearance quite similar to Wayne. George is quite a fan of his and was no doubt swayed by the honesty and forthrightness for which this commanding figure of a man stood. These were the same qualities of character that his father had raised George to value.

George and Helen could tell within a short time that Joyce had become interested in Rich. He seemed right for her. Lucy looked at no one other than Jimmy. Both girls had begun to discuss their plans for marrying. This would become a year for weddings and receptions. The women of the house were busy choosing silverware and china patterns. They were full with the excitement surrounding the preparations for such memorable occasions.

George had, by late fall of 1949, become use to people referring to him as the soon to be "Father of the Brides." Both of his lovely young daughters had found the young men they wanted to marry. George, of course, had teasingly, offered to pay five-hundred dollars to each of his daughters if they would promise to elope. They both firmly opted for a church wedding and they would be marrying within six months of each other. Joyce and Lucy choose their wedding dresses from bridal shops. Their bridesmaid's and flower girl's dresses were made by a seamstress. Joyce was the first to graduate from high school. She and Rich let it be known that they intended to marry in December of 1949, one year after Jimmy had introduced them.

Joyce May Shroyer married Richard Glenn Roberts on December 17, 1949.

Helen, reaching her full maturity, was a distinctive figure of loveliness and strength. Her days were filled with the responsibilities innate in helping her daughters plan for their weddings. She had also busied herself making baby clothes, as she was again carrying George's child. She glowed with the resplendence only a pregnant woman attains knowing she will soon bless her husband with the richness of a new life.

George, celebrating the birth of his new son, blew the air horns on his first Cadillac from the moment he pulled into the long curving driveway all the way down to the garage. Joyce, Lucy, and Georgie were all hoping for a boy and they were in the house awaiting the news. They knew when they heard the air horns that it must be a boy. George was just beaming when he came in with the news. Rex Gordon Shroyer was born April 5, 1950. George and Helen both had great admiration for Rex Clemens. This, no doubt, is why they choose to name their new son after him. "Little Rex" brought great joy to his parents and siblings. He was the child none had ever dreamed of.

Lucy married two months later. She recalls a special moment that occurred during her wedding. Mamye Fields, Johnny's five-year old daughter was Lucy's flower girl. George's sister Marie's little boy, Roger, was only four years old. He was the ring bearer. He'd been told to stand there just like Jimmy did during the wedding ceremony. When it came time for Jimmy to kiss the bride, Marie's Roger seeing this, leaned over and kissed Mamye in an innocent never-to-be-forgotten moment: A first kiss.

Lucy Ann Shroyer married James William Patterson on June 11, 1950.

The previous months' celebrations of two marriages and one birth made life seem great. George and Helen were also looking forward to becoming grandparents. Joyce and Rich were expecting their first child in October of 1950.

Early one morning in late August, George awoke with a strong premonition of an impending death. This was quite unusual for this man who had seen many tragedies in his day. This premonition was so overpowering he felt he had to do something to avert what he believed was a coming tragedy. He called the Corvallis Airport to ask his son-in-law Rich and his father not to take off in Doctor Faust's small plane. He was told that they had just left.

George knew Rich and his father were planning to fly to Seaside with a friend to make reservations for a deep-sea fishing trip. It wasn't more than a few hours before they received the fateful news of the plane crashing near Newberg. Reportedly, Harold Robert's told rescuers: "Please, take care of my son first he has his whole life ahead of him." Rich, 24, died soon after reaching the hospital. Harold, his father, was the only survivor. Richard Glenn Roberts died August 23, 1950, and is lying in the Oak Lawn Cemetery, Corvallis. This was only six weeks before the expected birth of their first child. Joyce was facing a devastation few young women deal with. Grief-ridden, she must endure the remaining weeks of her first pregnancy alone. Joyce returned to live with her parents while awaiting the birth of her child.

Joyce gave birth on October 8, 1950, to a daughter. She named her Kathy Glenn Roberts. George and Helen had become grandparents for the first time. Their second son Rex was six months old when Kathy was born. Now, there were two babies in the home they lovingly called "The Castle."

Mr. Hunter, Helen's father, suffered his first stroke soon after Lucy's wedding. The stroke left him nearly blind and barely able to get around therefore everything he did now took considerable effort. He still went about his chores but life was now much harder. George and Helen now had a young daughter who had recently lost her husband, two newborn infants and an elderly, feeble parent to care for. Helen bore the brunt of it but Joyce helped her care for her grandfather and the new babies. That year was especially difficult for them. George was busy working most of this time up at Detroit Dam, about one-hundred miles away, on the Western Slope of Oregon's Cascade Mountains.

The following spring, Mr. Hunter suffered a second stroke. While he was in the hospital he kept asking for his babies. The nurses thought it odd, that a ninety-year-old-man would be asking for his babies. They told Helen, when she came in, that they thought he was hallucinating. Helen told them; "Oh, no! He's asking to see the babies we have at home." What he meant was, he wanted to see Rex and Kathy. Joyce and Helen brought their babies in that evening and he died the next day. William George Hunter died April 20, 1951. The family's previous tragic loss magnified their pain. Not only had their grandfather had been a significant part of his grandchildren's lives, they also felt that they had lost a parent. Helen deeply mourned his parting. Although his death relieved his suffering, she selfishly wished him to continue living. He had been a source of comfort to her throughout her life and she would miss his very presence. Mr. Hunter was buried beside his wife, Rosa Nell Prather Hunter in the Wren Cemetery. George and Helen's baby girl lies there also.

> I remembered the last fun I had with Grandpa Hunter. One day after Grandpa's first stroke, while he was weak and nearly blind, I took his walking stick and replaced it with a heavier one. That night he complained of feeling very weak. Early next morning early before I left for work. I took the heavy rod away and put his own walking stick back in its regular place. That night when I got home I asked him; "How are you feeling today Dad?" He said; "Stronger, yesterday I could hardly lift my cane and today I can walk right along."

Helen had "The Castle" remodeled to make room for her growing family. Joyce was staying in the "Blue Room" with Kathy when she picked up the remains of her broken dream. That fall she returned to school and began taking business courses at Oregon State College. They decided Helen needed help, so she hired Francis who lived near by and she worked for Helen until Rex started first grade.

> While Mama was busy with the family, I was busy finishing up with the Detroit Dam. Mama and I bought an unfinished sawmill and about six-hundred-forty acres of timber fifteen miles up-river from Yachats. We called it the 'West Coast Lumber Company.' We were using three, single 671 GMC diesel engines for power. We cut sixty-thousand-board-feet per day in there with about eleven guys and we had Mama's brother, Bill run the mill. There was a real nice cook house and bunk house there, where some of the guy's stayed. Ernie was the wood's boss while we were logging that. Ray Tyson had the job of hauling that lumber into the Bennett Hart Planer in Philomath. Later, we hauled that to Bill's Planer there in Newport.

> Next, we brought our sawmill that cut eight-foot studs for the Detroit Dam site down here, and set it up in Philomath where we ran that for a while before we moved it over to Newport in 1952. When we set it up

in Newport, we made it longer since we needed it to cut longer lumber there along the coast. This mill was run by electric power and it could cut about thirty-five-thousand board feet per day with a crew of six. We set that up near the planer that Bill owned there in Newport.

One time I was needing some railroad spikes and the railroad foreman in Philomath didn't have any used ones. So, I got in my Cadillac and drove to Eugene to buy a box of new railroad spikes. Later that night, when I got home, it was real icy in our driveway – normally we didn't have much ice. Thought I'd get this box of spikes into my pickup so I'd have them with me the next morning when I took off for the woods. I was lifting this box of spikes from the trunk of the Cadillac. Instead of pulling the pickup's tail gate down like I should have, you know; I threw that two-hundred-pound box of spikes up over the side of the pickup. When I went to do that, one of my feet slid out from under me, and down I went, spikes and all. I landed on that hard cement, really wrecked my back that time.

The next morning I couldn't hardly stand to ride in my pickup. I had a lot of trouble getting around after that, couldn't do my work. It got so bad I couldn't even set in my pickup, and it kept getting worse. Finally I agreed to having back surgery, had that in December of 1952. I had to go to Portland for a spinal fusion on my lower back. When the doctors said I could get up and around again; I found out that setting in my pickup just killed me. So I bought myself a real comfortable car to ride in, a yellow 1952 Oldsmobile. It was a long time before I could set in my pickup again.

Seems to me that was about the time Lucy's first little girl was born. After my back operation, I couldn't go to work for some time. Lucy would come over to see me and to help Mama some. She'd bring 'Pammie' over with her and I'd hold her for Lucy while she helped Mama. She was a cute little thing and they were sure proud of her. They named her Pamela Sue Patterson, she was born April 21, 1952.

Since I'd broke my back I decided to form a partnership. Eugene Fenner and I had worked together over the years, we always called him Shorty. He'd worked under me when I was running things for Rex there at Eckman's Slough in the early Forties. Shorty was interested so we formed a partnership in the early Fifties. We called it "F & S Logging" for Fenner and Shroyer. Shorty was a good logger, he worked at Glenbrook, and they probably got more done than anyone in the country. One day when he was cat logging with the addition of the yarder, they sent out thirty-seven loads, he was a real good 'cat skinner.' These were 'Old Growth,' a few of the largest cuts would run eight-thousand board-feet with an average diameter over forty inches on the butt.

> **Shorty always drove the crummy back and forth each day to the logging site. In the summer he'd leave our shop in Philomath at three-thirty in the morning and arrive at the landing in time to start work at five o'clock. They'd work ten hours with a-half-hour off for lunch. In the crummy the guy's would get to wrestling, they were just having a good time, you know. Old Shorty, he'd threaten to stop and throw them all out. They played real hard and worked real hard.**

Another tragedy befell this family. Lucy and Jimmy's little baby girl became ill with a bronchial infection just short of her first birthday. Within a couple of hours after being admitted to the hospital her life ebbed away. Another marker was placed in Oak Lawn Cemetery. It reads, "To Our Little Angel Pamela Sue Patterson." George and Helen grieved greatly for Lucy and Jimmy's loss. No one could have predicted such tragedies would take place at a time when families normally experience the most joy in their lives.

Earlier that year, Charlie Hall started seeing Joyce. He was a navy veteran and a responsible and level-headed man. He was Rex Clemens' neighbor and he'd begun to show some interest in Joyce. One time Rex was giving George a bad time saying he thought Charlie was just a "passing whim" for Joyce. George said; "No, Charlie's having ideas about taking Joyce away from us." Rex said; "I'll eat crow if Joyce marries him."

Charlie seemed to have become the father that Joyce's little girl needed. He figured more prominently in Joyce's life as the year wore on. Joyce and Charlie announced that they would marry in the fall, on September 5, 1953. She loved him dearly and was happy to have a responsible and caring man to share her life with. They had a full life together and reared four children.

Sure enough, at the wedding reception George presented Rex with a "well-done crow" served on a silver platter. George had a little help from Georgie the night before for they'd gone out into the fields and shot it. At the reception, George surprised Rex and the guests when he proudly brought in "the crow." Everyone sure had a big laugh over that.

Soon after the wedding, George and Helen began searching for a cattle ranch they could invest in. One Sunday, Jimmy saw an advertisement in the *Oregonian* for a cattle ranch in Idaho. Helen called up to find out more about it, and was told that it was called the "Billy Creek Ranch." This beautiful Idaho ranch lies along the horseshoe bend of the Salmon River. This section of land lies near the confluence of the Salmon and Snake Rivers in Idaho; it contains about twenty-thousand-deeded acres and sixteen-thousand-leased acres. Cottonwood, Idaho, the closest small town, is about forty-miles away. When Helen and George first looked at this property, fall of 1953; Helen fell in love with it the first time she saw it.

The Shroyers' took possession of this property the first day of January in 1954. George and Helen took Georgie and three year old Rex as well as Lucy

and Jimmy with them for a visit to the Idaho property. George bought a four-wheel-drive power-wagon so they could get up the mountain on the unimproved road. That area was quite remote.

The "Billy Creek Ranch" is the single largest parcel of non-federal land along this Scenic River Corridor, the Nez Perce National Historical Trail runs through the property. In about 1840 Chief Joseph was chased by the U. S. Cavalry across this high prairie, down the benches and across the Salmon River some four-thousand-feet below. Today, these high mountain plains are known as the Joseph Plains. One of the cavalry is buried along the trail on a lower bench of land, this grave is listed in the *National Register of Historical Places.*

The surrounding landscape is high mountain desert. This property rises up some four-thousand feet from the river to the top of Joseph Plains. It's made up of three finger-like flats. The southern most flat, Rattlesnake Flat, aptly defines its physical formation. This flat is a rather narrow, zig-zagging flat extending out a couple of miles from its base. From its north facing side issues multiple springs, they form a small, clear flowing stream lying entirely on the Shroyer property – Billy Creek. The middle flat is rather large and provides the most forageable land on these flats. The northern most flat lies in a northerly direction and consists of multiple levels of small forageable parcels containing ten or fewer acres. Occasionally, steers become stranded on these parcels and they must be driven off. The cattlemen must dismount and walk their horses down the steep descending spine of rock formation that provides the only access to these small parcels. This rocky terrain is too steep for horses to carry a person's weight during the decent, they can only carry them up-out-of-there.

Originally, these steep slopes were covered with Old Growth Douglas Fir and Ponderosa Pine. It is home for herds of elk, three species of bear, white tailed deer, cougar and many small animals. It is a working cattle ranch with both a summer grazing area and a winter grazing area. This ranch winters their cattle on the fields lying eight-hundred-feet above sea level. The winds blow the snow that falls here off and the air is warmer so it soon melts in the fields below rattlesnake flat. Here is a herd of about five-hundred cows with about four-hundred-fifty calves. They winter on the lower bench next to the wild Salmon River as it forms the western most perimeter of the property. The wild grasses from these fields provide food for the winter cow-calf-operation, they winter here until about May. Then they are herded up to the grasslands about five-thousand-feet above sea level. The calves are kept as long yearlings. This means they are born in the spring, weaned in the fall and wintered separate from their mothers, then kept on grass all summer and sold in the fall as long yearlings, the majority are Herefords.

(LEFT) Mamie Fields at "Billy Creek" Ranch. (RIGHT) Rex Shroyer at the ranch.

When George and Helen first bought the ranch they continued to hire Dan and Ben Mater, the two hands that had been running it. Their responsibilities were to raise and harvest the hay, and look after the cattle; they were good cattlemen. The ranch paid for their groceries and allowed them several hundred dollars a month in wages. George and Helen have always taken a dim view of drinking. Helen knew those men couldn't be eating as many hams as they claimed they were. She found out that they were buying whiskey and charging them out as hams. Dan and Ben finally lost their jobs. After that, Helen always hired cow-hands who had families and were not drinkers.

CHAPTER 20

A SAWMILL IN BLUE RIVER CANADA

A Letter of Appreciation — by Lonie White

I first met George right after World War II. I wanted to work in the woods and George was willing to give a returning vet a chance. George influenced my life more than any man I have known. When George would come out to the job site he always had a big-old smile, and a good word for everyone. Seeing George was somewhat like seeing a little sunshine; George had a very positive attitude. I don't think I ever heard him say anything bad about anyone.

Working for George was a pleasure and he paid top wages. Working with George was sort of like a game, he expected a good day's work—you gave him a good day's work and he appreciated it. We enjoyed working for George; it was fun. We could see the result of our efforts every day and it just felt good. In turn, George took care of his employees. Anyone could go to him with a problem and he would listen and help in any way he could, it was a lot like "being family." I learned to admire George; he became my role model. Because of him, I became a better husband and father. I have tried to pass these principles on to my children; the quality of my life continues to improve.

The time we spent in "Blue River" was a real adventure. One of the things I remember most was that George always had fun, it was contagious. Every day was exciting with new and unexpected pleasures and challenges. In winter, the nights were beautiful with the moonlight reflecting on the fresh snow. It was exciting to feel the crunch of the snow underfoot and to hear it squeak with every step. It was so quiet—you couldn't hear anything but the still of the night and the sounds you made. We had all the natural beauty, the lakes, rivers and the Rocky Mountains. You would see deer, bear, moose, cougar and other animals, it was a wonderful place.

Starting before Thanksgiving, it would snow almost every day, and the snow would be so deep before spring you couldn't move. When spring break-up came, there would be water everywhere. The North Thompson, a relatively small river would become a raging torrent, bringing with it, tree stumps, limbs, and whatever else—down the river. It was a never ending battle to keep debris from destroying the bridges.

"Blue River" was a railroad town, they serviced every train that came in; they did that in the round house. There was a sandwich shop in the train station; we called it the Beanery. There was a small hotel where the railroad employees could stay overnight; it had a restaurant and a pub. There was a small pool hall across the street, and behind the hotel was the one room school. There was no garage or service station because most of the year you could not drive

anywhere due to snow or mud. If anyone wanted fuel they shipped it, in fifty-gallon drums.

From Vancouver, B. C., to Kamloops it was about two-hundred-sixty-five miles. Kamloops is a large town now; then, it had a population of ten-to-fifteen-thousand. They closed most of the stores every day for tea time. Blue River was a one-hundred-fifty-mile drive from Kamloops with only a few very small towns between. After the first fifty miles, it became a gravel and mud one way road.

Vavenby was about fifty-miles south of Blue River. There were many abandoned log cabins that were falling down; most of these were one single room, about twelve-to-fifteen-feet by twenty-feet. There were a few farms also; these showed the ingenuity those people had, they'd used whatever was available. Vavenby had real stores. There was a service station there and a grocery store with almost everything in it.

Avola was about twenty-five miles south of Blue River. The road from there on was unimproved dirt with ruts, slides and other surprises. I remember very well, one place where the trail went between a rock cliff and the North Thompson. It was only passable if the water was very low. The road was so narrow we had to use a plank to cross a short section a few feet long and directly above the North Thompson. Avola had one very small store where you could get a few staples, lamp oil, flour, etc. "The Diamond Lumber Company" had been there at one time.

After that first spring run-off, they had a crew putting down a corduroy road in a swamp, the mosquitoes were very bad. George came out to inspect the work and when he got back to the office, George asked to have this crew paid in over time. George had a lot of empathy for others.

I remember the first cook house we had; it had a little gas generator. I believe there were two ladies doing the cooking. It was dark when we got to the dinning table in the winter time, the food was wonderful. Almost anything you could desire was there, it was hot and well prepared. I also remember the times George was there, he would always eat with us. George would always want to be sure everyone had been in to eat. If anyone was missing, George wanted to know if they were all right.

Thanks again,

Lonie & Marge White

Memories -- By Gene Byler

It was fall of 1954 when we started out for Blue River. We were excited about moving to Canada to work on the sawmill for George. My wife Maxine and our kids went with me. We owned a Jeep station wagon then, we thought that could get us anywhere we needed to go. We had our five young children, all under school age, packed in there. We stayed over night in Kamloops, we only had one-hundred-fifty miles further to go, to get to Blue River. There had been good roads most of the way. By the time we reached Clear Water, seventy-five miles from Kamloops, the road had become a graveled one way road, thirty-five-miles later, we faced an almost impossible road. What should have been a three-hour drive, became an eleven-hour drive. We arrived late that night in Blue River on Thanksgiving Day of 1954, that was certainly a memorable trip, we used train service after that. Four passenger trains came through daily, two going east and two going west.

Blue River lays north of the United States, about two-hundred-fifty miles above the Canadian border and one-hundred--miles northeast of Kamloops in British Columbia. That's where George dreamed of building a steam-powered sawmill. Bill Hunter, my father-in-law, was hired as manager for the sawmill in Canada and he hired me as head sawyer.

- - - - - - - - -

George said; Ruby Moug from Oregon ran a real good cook house for the whole outfit there in Canada. One time a couple of French Canadians were hired on. That first night they came in for dinner one of those guy's thought that Ruby's seventeen-year-old daughter looked real nice. He whispered something in her ear when she brought the food around. She turned bright red and ran off to the kitchen where her mother was. Ruby came out of there with a great big kitchen knife. Boy! Those guy's took out of there, and they never went back in the cook house again. I talked to those French Canadians later about it, they said; "French Canadians have different ways, we don't know the—American ways."

This was to be a steam powered mill, Bill Hunter and a crew of my men moved there from Oregon. John Nauert, or Slim was the planer operator, he had worked for us at Newport. Dick Elee had been our bookkeeper at Newport; he moved up to Canada to do the book-keeping there. Claude Baker was our millwright there in Canada, he'd worked for us before. First, as a supervisor and general mech-anic on heavy-equipment, trucks, cats, donkeys, etc. on the Detroit Dam clearing project. Then he began working at the sawmill in Lobster Valley on a regular basis and did some work at Yachats and Newport later.

George Shroyer built a bridge over Canada's Thompson River

Bill hired some more men from Canada to help build the mill, we built that mill to handle one-hundred-thousand board-feet per day. The main building was sixty-feet by two-hundred-feet. There was a planer shed off to one side and a boiler room.

We'd bought two large pieces of equipment, they were delivered directly from the factory by rail car. The gang saw cost us about thirty-five-thousand and the new mobile crane cost about fifty-thousand. Two booms came with the crane; one was a sixty-foot-lattice boom and the other was a fifty-foot-heal boom for loading logs in the woods. That boom was large enough to cover a one-hundred-foot diameter of rough ground. One time, I yelled up to the crane operator as he was unloading the gang saw from the flat car. I said; 'If it tips over, crash the gang saw—not the crane'! You can keep a mill running with a crane, can't do much without logs.

We built this mill by fabricating the rafters and ceiling joists on the ground. There was only four-to-six-feet between the rafters. From the ground level, the floor was ten-feet high. We'd used eight-inch-square posts from the ground-level up. These were capped with off-set eight-by-eight timbers from twelve-to-twenty-four-feet long; these formed the floor joists. Then, we used that new mobile crane with its sixty-foot-lattice boom. We lifted the fabricated rafters and ceiling joists and set them in place on top of that floor, Gene did that.

When we were putting the roof on the mill, it was winter time. One particular day, it was raining and freezing, we wanted to get the metal on the roof. Gene and another guy went up there to work on it, wasn't long before they came down. They said that it was too slippery for them. Bill was standing there with me when they came down, we figured we'd finish that job up. So, we got up there on that roof. Boy! That darn roof was slippery, we came back down real quick. We figured we'd better get down before we slid right-off there.

The mill yard was deep with snow; every once in a while, I'd take a seven-foot bucket-loader, and shove the snow off. Johnny Nauert, the planer foreman was out in the yard. There was a large twenty-foot-long culvert, I plugged up one end of it with the loader. Then, I went over and told Johnny that darn culvert was plugged-up and I wanted to use it. He got a little shovel and got in there, the minute he got inside, I hurried fast and plugged-up that end too. Then, I started to roll it; I could just hear him in that culvert cussing-like-crazy, so I'd roll it again. Well, he finally ran out of cuss words and I let him out.

Every morning, this old guy would come around and take about an hour to walk around the mill. He was always walking close and looking at everything. There was a big berm of snow along the side of the mill, where the snow fell off the roof when it melted. He liked to walk there because it was real compact. One morning I said; "Better watch it! It won't be long before the snow on that roof will come down." He said; "No! No! I'm all right!" Not long after that, I heard it start to slide off the roof. I looked out and there was nothing showing but his feet! A bunch of guys ran over to shovel him out. After that he stayed back – way-back.

One of the first things they did, was pour a foundation for the gang saw. First they built the form for it, this was an eight-foot cube. Next, they poured the cement they'd mixed with calcium-chloride; this caused the freshly mixed cement to become warm, allowing it to cure without freezing. It was necessary to add this chemical to cement in such a cold climate. It took about one-dozen large anchor-bolts to hold the gang saw in place, these steel bolts had a two-inch diameter. They placed each of the thirty-inch anchor-bolts into the cement twenty-four-inches deep, leaving six-inches remaining above the cement. Alignment with the holes in the base of the gang saw was real important. This gang saw's base was about four-feet by six-feet. Something kinda like a crank shaft holds the blades. The blades on that gang saw were about four-feet long, ten-inches wide and about three-sixteenths of an inch thick. The teeth are like those in a rip saw and it cuts as it pulls the blades down, they pull away from the wood as they return to their cutting position. A mill can produce two-to-three times as much lumber with one of these gang saws.

Next, they poured a big cement-slab for the dutch oven's foundation, that would be big enough for a man to walk around in. We needed it built that first winter, it would act as the firebox for the boiler. In the early days, people believed you could only have upright boilers. Inside the dutch oven they built a huge horizontal boiler. It was about twenty-feet long, ten-feet high, and eight-feet wide. It contained about one-hundred-to-one-hundred-fifty pipes that lay length-wise. They were rolled in at each end onto a steel plate one and a quarter-inches thick.

Each pipe had about a six-inch diameter and was formed of malleable steel. The distance between pipes was about one-half-inch. The hot air from the dutch oven was drawn through these to make steam. The draft was created by the addition of a fifty-foot-metal smoke stack on the dutch oven. At the top of the boiler was a two-foot bubble to take the steam off. This steam was blown through pipes to the four steam engines needed to run the mill.

As the ground heaved during the spring thaw, big cracks opened up in the foundation of the dutch oven. It had so many large cracks in the foundation that the air flowed in constantly, we couldn't maintain the temperatures needed. I was in Canada when we first fired that up. That old boiler really started boiling, when it built up enough pressure to put-off steam it began to shake continually. When it did, the brick work began to fall apart; it soon had so many large cracks, that the air flowed in constantly. You couldn't maintain temperatures needed to generate steam, we sure needed that steam to power the mill. We had four big steam engines depending on it. Those big twin-cylinders we expected to run with, were real powerful. The cylinders were off-set so they ran real nice. The cylinders had a twelve-inch-diameter piston with an eighteen-inch stroke; the longer the stroke, the more powerful the engine.

When the guys would go to work running that head-rig saw the power would only last twenty minutes or so. Then, they would have to wait for about fifteen minutes for the boiler to build up steam again, that would never work. We never were able to succeed with that mill, we tried diesel engines and stuff, none of those worked very well. We were counting on that steam power for lots of things.

This resulted in a mill that could not produce sufficiently to pay the workers, it was an impossible situation. The Canadian government's boiler inspector came to check it out. In those days they ran a cold water test on boilers by filling them completely with water. Next, they would close the valves leading from the tank, they would then heat the water some. When water is heated, it expands -- this makes any small leak show up at once. If it leaks at three-hundred-pounds per-square-inch they require you to repair it by welding the seams, replacing pipes and valves, etc. Then, they would allow you to run at the normal operating pressure of two-hundred-fifty pounds per-square-inch. The burner was fueled by sawdust and planer shavings from the mill; these were blown into the cyclone and dropped into a bin. From there, they were transferred into the boiler room.

The dutch oven was condemned. So, the men turned their energies toward developing alternative power for the sawmill. They utilized diesel engines to power the mill. They had a Cleveland-five-hundred-horse-power generator, that was used to generate electricity for smaller motors and electric lights. Another, ran the head rig and gang saw. We had two, General Motor's diesel engines to run the edger and planer, they each provided two-hundred-horse-power.

The logs from the pond were literally, half-frozen. The part above the water was frozen-solid; while the side in the water was not-frozen. Three or four months each winter, they were dealing with half-frozen logs, saws didn't work well on these. This was hard on equipment and on the sawyer, it was extremely difficult to cut boards out of this timber.

They had originally expected to use heat from their large boiler to warm this shallow pond. The initial boiler was critically flawed leaving them with severely limited means of warming the pond. When faced with this unexpected problem they tried to remedy it by circulating the water. They blew air through many tubes running from their air compressors. They thought, if they could keep it moving, it might not freeze.

> Gene would go out to the pond to check the tubes we were using to blow air into the pond. He'd find the ends covered with globs of ice. That wasn't doing any good. We couldn't circulate the water in the pond that way. If it had been deeper, the ground temperature would have heated it some.
>
> One morning when we went out to get some work done, a six-inch steel shaft broke. Nobody had touched it. After that, when it got real cold I'd say 'Don't look at it—it'll break.' That's real cold I'd say! If it was a real cold day, it would get down as low as fifty-to-fifty-eight-degrees below zero F. One time it stayed that way for ten days. First time that happened, we weren't prepared for weather like that, we tried starting a cat. It wouldn't turn over, so we built fires around it to warm it, we finally got that cat started. Then, we realized it wasn't going to move; the tracks were frozen into the ground.
>
> Some of the workers were real hard on equipment. Another time, I walked by and heard this engine squeaking and I said; "What's

wrong with that engine?" The guy running it said; "Oh, nothing, it just sounds that way." So, I took a look at it myself, there was no oil in it at all, that would completely ruin an engine!

The logs dumped into the mill pond and were transported by a side delivery log-haul, the teeth along the transport, dug into the log carrying it up out of the pond. We used a skyline across the mill yard to move logs from the log dump into the mill. Skylines are held up by two poles in the ground, these are held in place by three or four cables from thirty to forty-feet long radiating out from each pole and staked into the ground.

George had never owned a big gang saw before he bought one for that mill in Canada. The teeth of this saw are similar to those in a rip saw, it cuts as it pulls down, it could cut twenty-four by twenty-four-inch beams. The saw blades on the gang saw moved up and down three-hundred-fifty times a minute as did the crankshaft driving them. The large gang saw always cut several boards at one time.

They had a double head rig with fifty-two-inch circular saws. The first pass takes a slab off, this goes into the burner. The remaining log is cut into smaller cants to feed the gang saw. These cants ran anywhere from four-inches-to-twelve-inches thick and up-to-twenty-four-inches wide. The gang saw then slices each cant into boards of multiple dimensions beginning with a, two by four-to-a-two by twelve or a three by four-to-a-three by twelve etc. The double head rig George had up there, could cut four twelve by twelve's, at one time. This was a common beam marketed in those days. Some of the boards from the head rig and the gang saw had to go to the edger to be re-manufactured.

The head sawyer's job is to operate the double head rig, this is the most critical position in the mill. The head sawyer decides what size cut the logs can take and what dimensions to cut out of each log. They had an order board that was put up for the sawyer and edger. This told them what dimensions were needed for the current orders. They would often be working on seven or eight orders at one time. The head sawyer instantly determines what he is going to do with the log. This requires him first, to decide whether there is any decay that needs to come off. Then, he must determine the best utilization of the remaining log based on the orders that need to be filled. In the older mill in Yachats, the sawyer and the man on the carriage communicated with hand signals. In the mill in Canada, the head sawyer ran the carriage by remote control, there was no one riding on the carriage.

One time, right after we got the steam in the mill working Gene figured he ought to test the new whistle he'd installed. He wanted to make sure it worked right so he pulled on that cord, it worked all right. That whistle got stuck and wouldn't stop. The men from the railroad station came runnin' up there to see what the trouble was. Took quite some time before he could get that stopped, we found out

that was the "signal for trouble" at the round-house for the railroad station.

Most of our orders came from England and other overseas markets; some orders came from New York. The Canadian National Railroad had a spur rail into the mill. We loaded pallets containing orders of lumber onto train cars with a machine we called a straddle bug. It had large wheels with stilts attached to them and we drove that around the yard. You would position it above a stack of lumber, pick that up and drive it over to the loading platform. Sometimes we used fork lifts to load the train cars.

We did our own logging for that sawmill, logging there was very difficult due to the weather. We would fall, buck and cold deck the logs right there in the woods during the spring and fall, you could only truck them in during the summer or winter months, when the roads were frozen. We usually did that in December, January, and February. We'd truck enough of those logs into the mill to get us through the spring since after the first spring rains, the roads were too soft to truck logs. Summers were real good for trucking beginning in June and on through July and August. We trucked our logs into the mill during those summer months before the fall rains began.

We had to build lots of roads since it hadn't been logged much before. We even built two bridges each about two-hundred-feet long, over the North Thompson River. We bought a pile driver for the crane, and we put four pilings in a row and each row of pilings was about ten or fifteen-feet ahead of the other row. We drove those pilings about ten-to-twenty-feet into the mud and left them up about eight-feet above the water level. We capped the pilings with some twelve-inch timbers. We drilled down into the timbers and spiked them into the pilings below. On top of the timbers we laid four logs running parallel, these were up-to one-hundred-feet long. We laid twelve-foot stringers across those logs. That made a real sturdy bridge.

We had the right to log up to sixty-square miles of timber but we didn't get much of that logged, we logged mostly close in to the mill. We were cat logging. We cut lots of forty-eight-inch-diameter Engleman Spruce; it was real nice. We sold this as pine. We also cut some Douglas Fir, a mountain species of good quality. There was some hemlock on west-facing slopes, most of the cedar was down by the river. It lay in there at an elevation of about two-thousand-six-hundred feet. Some of it was one-thousand-years old and was falling over. Most of it had an eight-foot diameter with a four-foot-diameter hole in the center of it, that would have made good clear cedar siding. Someone offered me a-quarter-of-a-million for it. Wished I'd sold it to him!

> One of the areas, we were logging, was known as the Red Sands area; it was a swampy area about two or three miles out of town. It lay near Mudd Lake, between the mill and the mountains. We built a "cord-uroy road" there so we would have access to the woods; we were able to build only two-to-three-hundred feet a day. We did most of that with a crane and we had a tong for moving the logs into place. Then we covered the logs with about two or three feet of gravel. We built about a mile of corduroy road. A "corduroy road" will last a long time if the logs are always wet or under water; it preserves them.
>
> Last time Gene came to visit he said; "George, I went back to Blue River a year ago; and found the old corduroy-road you built in that swampy area. Also, one of the bridges you built over the North Thompson River on the road to Red Sands, is still being used."

George spent over three years trying to get the mill in Canada to run thus Helen was often alone with Georgie and Rex. She no longer had her father with her for companionship, and both girls were married. Helen and George thought having a grandparent living with them again, would be good for the boys, and Helen got along well with George's mother. They planned on having her move in with them; she was going to stay in the "Blue Room," it would be a nice room for her. They needed to make a trip to the Idaho Ranch before moving her in. While they were there, the Taylors sent someone to give them the message that there had been a death in the family. They didn't tell them who died; so George and Helen feared it was one of the girls. After driving two hours to the nearest phone, they were told it was George's mother. She died of a heart attack while still living with Marie and her husband.

Mabel Louise Hatch Shroyer, 78, died November 2, 1956.

CHAPTER 21.

FINANCIAL BATTLES FOUGHT OVER CANADA

George and Helen thought a sawmill in Canada would be a good investment and George spent three tedious years trying to make this dream a reality. However, this venture would prove to be their financial undoing. They now were facing the worst financial difficulties of their lifetime.

> **Kenneth Watkins helped us find that property in Blue River, Canada; there sure was some nice timber up there. We flew up there to look at it early in 1954; the plane to Seattle was pretty big. When we left Seattle for Kamloops we took off in a small plane that carried about a dozen passengers, some were movie stars and they were flying in for a vacation.**

The expanse of land and the size of the timber reminded George of logging in the early days along the Oregon Coast. Never, since that time, had he seen trees of this abundance; the majority were Spruce and Fir, about thirty-inches on the butt. There were some large Red Cedar, from ten-to-twelve feet on the butt, growing along the river. This seemed like a logger's dream, plenty of timber and an inexhaustible market—all he needed was a steam-powered mill, what more could an investor ask for.

> **We bought the Blue River land and timber rights from Len Jones; we paid one-hundred-thousand-dollars down on it. Kenneth and Mrs. Wolf were there when Mama and I signed the agreement; we signed that on February 15, 1954. Sixty licenses or leased rights for timber came with the land.**

Each license is equal to a section here in the U. S. and each section equals one square mile of land. Therefore, they had the right to log sixty-square miles of land. George and the crew he'd hired were in a foreign territory where the rules were different, these rules were established by the Canadian government. Shroyers hired Barney Johnson as their Canadian engineer, they also hired a Canadian lawyer named Hamilton Reed. These men were responsible for the legalities relating to the licenses on their timber holdings. They were each paid four thousand a month during the entire time the Shroyers were operating in Canada. This expenditure reflected the monthly reports and other required legal documents generated for the Canadian government.

Shortly after George and Helen returned home from their trip to Canada they quite unexpectedly received a bill from Mrs. Wolf. This bill was a finders-fee for thirty thousand on a one-million dollar parcel owned by Len Jones who had agreed to pay a commission to Mrs. Wolf for finding a buyer. George asked Mrs. Wolf why she sent him a bill. She replied that Kenneth signed an agreement awarding her a fee for locating the land and timber in Canada if Shroyers bought it.

George and Helen maintained they had not given Kenneth permission to sign the agreement with Mrs. Wolf. She sued Shroyers later in 1954 to collect her finder's fee. The suit went for a couple of years with the judge eventually ruling in Mrs. Wolf's favor. The court required the Shroyers to pay Mrs. Wolf's finder's fee as well as all attorneys' fees. This cost the Shroyers eighty-thousand dollars. The attorneys James (J. K.) Weatherford and Orval Thompson represented Mrs. Wolf. Later, Shroyers hired them and they became friends. George always said, "That was a real expensive way to make good friends." Never, for a moment did George and Helen consider the cold as a major threat to this endeavor. The weather had been the worst thing there in Canada. From the very beginning there were hurdles that would prove impossible to surmount. Distance and unfamiliarity with the extremely cold weather played a big part in dooming this undertaking.

George and his crew were unprepared to deal with the Canadians' attitudes and expectations which contributed significantly to the undoing of their venture. Helen told of wrong doings that were taking place. The Canadians they were doing business with, had an attitude that could be summed up as "Shroyers are rich Americans." These Canadians believed they should get every thing they could, from them. After sinking all that time and effort into the sawmill in Canada, Shroyers found themselves one-million dollars in debt. So, George closed it down. Through all of his difficulties, George never wavered from the way his father raised him.

> Some people thought I should take out bankruptcy but my "Old Man" didn't raise me that way. We mortgaged the ranch at Missouri Bend. That was our first ranch; we bought that from Dr. Blower. We also sold the timber off the Idaho Ranch. We contracted that out to an outfit that did helicopter logging, they put in three helicopter landing pads up there. That paid off our most urgent debts. Things were pretty bad!

George was broke when he returned home from Canada in the spring of 1957. They still had to make the last payment on their spectacular Idaho Ranch. Life was difficult for George and Helen with the loss of the sawmill operation in Canada. After this financial tragedy, George had come back to Helen completely exhausted. He began helping her in the garden. This was the first time ever for him to offer to work in Helen's garden.

> When I first got back from Canada, our final payment of three-thousand-five-hundred dollars was due on the Idaho Ranch and Mr. Youngberg sent us a notice to pay that. We tried to call him to ask if we could pay that last payment "a little later." We'd agreed to pay him a-quarter-of-a-million dollars for the Idaho Ranch. Mama and I found there was a late payment clause in the contract that said he'd get the ranch back, if we were ever late with a payment. He figured if we were late making that last payment of three-thousand-five-hun-

> dred dollars, he'd have the Idaho Ranch back and he'd still have the two-hundred-forty-six-thousand-five-hundred dollars we had already paid on it. 'That would be a real big bonus'! I'd say.
>
> Mr. Youngberg wouldn't answer his phone. So, Mama and I drove up there near Portland to see him. When he answered the door he wasn't expecting to see us. Mama went right in and said; 'Mr. Youngberg, what do we owe you to pay-off the Idaho Ranch'? He told her; and she pulled out the checkbook and wrote him a check for 'Payment In Full.

George had lost his reserve of inner strength. These last grueling months in Canada had exhausted him. He only smiled at Helen recalling the pithiness behind her youthful, terse little smile. Her spunk was one of the things he'd loved from the moment they'd met. Helen had saved their investment by relying on the grit she was made of.

> Mama and I went right back to Philomath. Mama went into the bank here and said; 'I've written a check for three-thousand-five-hundred dollars to save the Idaho Ranch. We don't have any money left—do whatever it takes. Just cover that check.'

Georgie recalls watching the bank appraise their home; Shroyers had to mortgage their "Home Place" to save their ranch. George and Helen spent the following summer working on their ranch in Idaho. They couldn't afford to pay anyone to herd the cattle or put up the hay on the place so it was their job that summer. That was the only way they could make do. Georgie, a teenager, six-year old Rex and Mayme Fields, his nanny, went with them. Mayme went along to help keep track of Rex while the others worked. They were filled with anticipation and excitement at the prospect of spending the whole summer up there on the mountain. They worked hard that summer building a hay barn, putting in the hay, and caring for the cattle, this was a good summer for them. Just being together as a family helped heal their wounds. This was surely a summer to remember.

George, now having the time to reflect on the failure in Canada, soon realized he still had many things to be thankful for. He had a small income from the Fenner & Shroyer Logging Company he had started with Shorty and they still owned their beautiful Idaho Ranch. This piece of land was the most spectacular setting anyone could ask for. George felt particularly thankful that Helen had stayed firm in her decision not to sell the Idaho Ranch to help save Canada. Today George says 'The summer we spent in Idaho was probably the most real fun I ever had.'

In the calm following this traumatic experience George found more contentment and enjoyment there at the Idaho Ranch than he would ever experience again in his entire life. While working in the fields during those warm summer days he felt serene. Working closely with those he loved on this land with the spectacular view was comforting. The joy of this summer forever locked in his memory; he would continually yearn to return there to live out the days of his life.

When they got back from the Idaho Ranch that fall, the table was covered with enormous bills George and Helen could not begin to pay. George was sorting the bills, the most critical bill was one from the bank for thirty-thousand dollars. There was a second envelope from the IRS, seemed to be a refund check for thirty-three-thousand dollars. This was almost exactly the amount they owed. George could not believe the government had sent him money. He asked Helen to look at it and tell him if it really was a check. Sure enough! It was a refund of overpaid taxes from the previous year – too good to be true. All the other envelopes opened that morning had been bills. Lady luck was smiling on George.

Mrs. Wolf and Mr. Parks met at the Portland airport restaurant for lunch. Mr. Parks was a banker who was trying to move into the timber industry. Mrs. Wolf had been working with him. George and Helen had signed a contract with Mr. Parks; he was offering to buy the Blue River property in Canada. Mr. Hickman, George's insurance agent from Eugene, was a good family friend. He was in the Portland airport restaurant waiting for a flight when he overheard Wolf and Parks planning to force Shroyers into bankruptcy. Then, they would take everything the Shroyers owned. Apparently, the contract George and Helen had signed with Mr. Parks included a hidden clause saying; they had to pay all debt on their property before Mr. Parks had to pay Shroyers any money at all. Selling to Parks surely would have forced George and Helen into bankruptcy. Len Jones, the original owner, had signed an agreement allowing them to sell the land and timber leases without first receiving "payment in full." Shroyers would not clear any money from the sale, it would only bring the lease payments up-to-date and pay-off Len Jones.

Mr. Hickman told George about the conversation he'd overheard, then he said; "George you are like a 'lamb' whereas; I'm like a 'wolf'! You just leave those guys to me. I'll take care of them, I have a two-fold detainment plan. First, I will trick Mrs. Wolf and Mr. Parks into believing I have a buyer who will pay a significantly higher price for the Blue River property. Second, in case I'm not able to sufficiently detain them, you and Helen go to the Corvallis Airport, saying, 'You are going to wait for Mr. Parks to fly in.'"

Mr. Hickman called Mrs. Wolf indicating he had a buyer for the Blue River property and he'd heard she was handling the current contract for Mr. Parks. Mrs. Wolf said: "The time limit on Mr. Parks' contract with Shroyer's is almost up." Mr. Hickman suggested Mrs. Wolf bring Mr. Parks over to his office there in Eugene where they could look over the offering. Since Mr. Hickman knew his way to the Shroyer's home and it was somewhat difficult to locate; he said, "I'll be happy to take you out there afterwards." Mrs. Wolf agreed to the arrangement and thanked him.

Mrs. Wolf and Mr. Parks arrived at Hickman's office at the designated time (Parks contract with George and Helen would expire at midnight, that night). They studied the contract Hickman presented them with carefully; it seemed to be a substantial offering. Mr. Hickman started talking quite a bit. Finally, Mr. Parks said; "Let's get this contract signed." Mr. Hickman quickly found a pen. It didn't work, apologizing profusely, he disappeared into the other office reappearing after an interminable time. He said; "This pen ought to do it"—of course that pen wouldn't work either. Mr. Hickman, showing great surprise and irritability, ran out of the room. He managed to do this a couple of times before he finally came up with the right pen. By this time, they were beginning to doubt his sincerity. Their suspicions of pretense on his part were roused, and they were trying desperately to maintain their composure. Mr. Hickman, while in the other office, had in fact called George at home to let him know, shortly, they would be on their way to his home.

George and Helen now put their disappearing plan into action. They would soon leave the house pretending they were expecting Mrs. Wolf and Mr. Parks to meet them at the Corvallis Airport. Before leaving, George called his next door neighbor, Harold Roberts, he and his wife Hulda came right over. They had agreed to await the arrival of Mrs. Wolf and Mr. Parks at George and Helen's home, thus serving as witnesses to their late arrival. Harold brought with him a precisely set time piece. Mr. Hickman successfully detained Mrs. Wolf and Mr. Parks. They were eleven minutes late according to Mr. Roberts' time piece. With a great deal of assistance from friends, George and Helen barely escaped the forced bankruptcy trap set for them by Mrs. Wolf and Mr. Parks.

> **Helen said; "Len Jones was as-good-as the others-were-bad", he was a real nice guy. He could have foreclosed on us; I'm sure. He never even mentioned doing that. We got to know him well during those years.**
>
> **George said; Len Jones was always helping his friends. He owned several service stations, his friends would want to own one also. So, he'd give them some money to get a start and before long, they'd need to borrow again, so he'd give them more money. Pretty soon, he'd own that service station too.**

George felt it was more honorable to struggle in the face of poverty than to take the easy way out and file for bankruptcy. A full ten years would past, before George was finally able to make the final payment to his creditors. He never forgot the pain of that experience; it always reminded him of Rex Clemens saying; "You should always stay where you know the people and the people know you."

> **We were returning from our 1980 trip to Quebec, Canada. We wanted to see Len Jones again while we were in the area. Len flew from**

Victoria, B. C. to meet us at the airport in Seattle. He told us he was dying of cancer and wouldn't be seeing us again. He told us that Blue River was a big financial loss to him also. He said it had changed hands a couple of times after we lost it. The other investors had also lost a great deal of money there. That land and timber in Blue River, Canada seemed to have a jinx on it. Blue River broke everyone that owned it.

Helen said: 'Seeing Len again reminded me of one of our first winter trips to Canada when, George and Len hid their eyes when we were trying to land. Once when we flew back to Vancouver, our plane circled the airport for sometime. The runway had iced up, it had rained, then frozen into a sheet of ice. It was kind of funny, here were these two-full-grown men with their hands over their eyes. They had them closed tight, as if that was going to keep us from crashing. They couldn't stand to look out the window while we were landing.'

George and Helen took a trip to visit Huntington (town) in Huntington (county), Quebec (providence), Canada.

Helen said: 'This is where Grandpa Hunter, my father, was born in 1861. Both my Grandfather and Grandmother Hunter are buried there as well as my great-grandfather and great-grandmother. I was named after my great-grandmother, Helen Hunter as well as my Aunt, Helen Hunter. We found my second cousin, Guy Hunter, living in Manchester, New Hampshire. There are not many people with the name Guy Hunter, we were able to find him in the local phone directory when we visited New England later that month.

'My ancestors were important people in Scotland, they were merchants and owned a big shipping line. Stocks were sold on the street corners and the money was given to the Captain of the ship to invest. If the ship didn't make money the owners became debtors and they would literally be quartered, in those days. Three of my ancestors ships were sunk by Napoleon during the war of 1812. When the news first reached them they took their jewelry and went immediately to the harbor where they bought passage on the first ship setting sail. The ship they'd boarded happened to be headed to Canada, they barely escaped this unreasonable punishment. After arriving in Canada, they went inland where they quickly established themselves, becoming as respected as they had been in Scotland. Both the town and county they settled in were named after them.'

CHAPTER 22.

SHROYER LOGGING BEGINS AGAIN

George, now fifty years old, was starting over again. That fall George and Helen returned home from their Idaho Ranch with Georgie, Rex and Mayme. George was heavy with the realization that he would have the grievous task of beginning again. When George returned from Canada, Rex wouldn't give him a logging contract. Charlie, George's son-in-law, had been sub-contracting with the Fenner and Shroyer partnership while George was in Canada. George figured Charlie might join him in a logging venture. So, George invited Charlie to work with him. They didn't have much in the way of equipment but figured they could put something together. George owned one logging truck and a D-7 cat. This cat was the only piece of equipment he'd been able to salvage from Canada. Charlie had a couple of logging trucks and his cat. Together, George and he bought a four-thousand-dollar yarder. With this limited equipment, they started logging together in 1957 and continued to work together through 1959.

They first logged together under a contract with Willamette Valley Lumber Company. This operation was up Yew Creek where they were going to be cat logging in a very dangerous place as one particular piece had a sixty-percent grade. George knew the only way they could get this timber out economically would be to use the "yo-yo method" of yarding. Few logging operations would attempt yarding this with a cat for in most situations it would not be economical or practical.

The term "yo-yo yarding" describes a way in which two cats work together to bring logs out of a steep canyon. George bought all new drum lines since the job was in this dangerous area. George did not want to take a chance with old lines that could cause accidents. Especially since the young man he would be lowering down the mountain side was his daughter's husband. Although still quite young, she had already lost a husband to death. George prided himself on having safe rigging. He would always say: "Never had a man working for him been killed due to poor rigging." George considered human error the greatest killer of all.

George stayed above the canyon on the road running his D-7 Caterpillar. Slowly, he drummed Charlie on his TD-18 cat, about one-hundred and fifty-feet, down the steep canyon. There, Doc Hamm was working to choker the next big log onto Charlie's drum line. Afterward, Charlie drummed the log to his cat. Next, George would drum Charlie and his cat with the log, to the road above the canyon. Their work went well for each had great respect for the other's character and skills. They'd hired good men for their crew and treated them well. They were both concerned about keeping their men and their equipment in good order.

When we were working up there at Blue Mountain in 1958, I was setting chokers and whistle punking at the same time. We had the BU-50 yarder running and each time a log was going in, I'd climb on this stump to look the area over. We were planning where to build the road to the next landing. While I was waiting, I was picking out the stumps to avoid. I had my mind on something else, you know.

The main line was hanging up on a root-wad and I'd sent in the signal to skin it back. Whenever the line came close I would jump off the stump. Two times this happened and I'd jumped off, just in time. The third time it hung up on that root-wad the main line swung over and hit me on the side of the head. It knocked me off into the place where the log was coming along. I'd 'pass-out', then 'come-to', just before it got to me. I remember thinking; 'Why couldn't I have stayed out just a little longer.' I could see it was going to run over me and kill me. I couldn't reach my whistle. I was lying on it. Some trees were real close to me and I was barely able to roll over behind them. I can remember that just as well as if it happened today.

Charlie was standing on the landing and saw it happening. He ran over and helped me in and brought me home. Mama took me to the doctor and he cleaned my head, sewed it up and sent me home. I went back to work the next morning. This accident started all the problems with my neck which always bothered me after that. Guess I was lucky! Charlie said: 'That would have killed anybody else.' Charlie and I worked together a few years, then Charlie went back to his own trucking business and I continued logging. Charlie had started Hall Trucking in 1946.

The next spring, George landed a contract with Willamette Valley Logging Company for Shroyer Logging. He logged the south side of Prairie Mountain, near Glenbrook. Shorty Fenner and George did a trade through F. & S. Shorty took full ownership of their trucks and in return, gave George the yarder. George now had a high-lead operation going again.

Shorty Fenner was running F & S while I was busy in Canada, and he continued running that after I came back. F & S worked for Pope and Talbot, logging around Glenbrook and Dallas.

Sam St. Clair worked for us around Glenbrook, he was reminding me about that darn spar-tree that they pulled the top out of. Bud Carey was running the yarder, Merle Gee was running the loader, and the choker setters were Bob Fenner, Shorty's brother and Dale Breazeal. They were bringing up an eight-thousand-board-foot log that was getting hooked-up on some big stumps. It wouldn't pull. The whole crew was around when the top of that spar-tree came out, from the buckle guys to the top-guys. Sam remembered that well. He said, 'I was settin' tongs loading Bill Rampenthal's truck when that happened. The guy-lines were coming fast as they could come, and Old Bill and I were running fast as we could – we weren't stopping. The guy lines were hitting all around, it was like running through an 'egg-beater'. They cut about two-inches deep where they were hittin' the ground. I had my dinner-pail on the head-block of the loading machine and it cut it into pieces and all I ended up with was a spoon that looked like a bracelet. My thermos bottle was setting there and you could just see where the guy lines had picked it up. Old Cool Bud just locked up the yarder and laid underneath the two running boards. Meryl, he just jumped under the loading-machine and those drums looked like back-lashes in a big fishing-reel. That could have killed one of us. If we'd been "rider-blocking," that wouldn't have happened.'

A "raised tree" is one that has already been cut down and limbed. It is yarded to the landing where it is needed as a spar-tree for a high-lead-logging operation. Next, half-a-dozen plates are added, and the guy-lines are attached. It is then raised, using the mechanical advantage of blocks and cables and the guy-lines are then attached to the surrounding stumps stabilizing the raised-tree.

We had this "raised tree" at Glenbrook we wanted to take down. I told one of the older guys to go over there and take that tree down. He had a kid down there helping him and they got to talking. They started at one stump tearing the lines down, and they just went right around that darn tree, you know. They tore three of those down on one side—and it fell right over the top of the yarder. Well, that darn guy, he knew better. Heck, he'd raised half of them, he might have raised this one; we never let him forget that, he was pretty embarrassed. For a long time after that he was wantin' to go back to Missouri.

About that time George was working up on Blue Mountain, and Georgie was helping him set chokers. George Newman, the guy we called "Punk," was above them and off to one side falling thirty-inch-diameter timbers. They were watching him and saw one of those trees hit, bounce, and come down the hill. There was a bench above them, and they thought the tree went-off to the side, but it didn't. Georgie yelled; just as he saw it coming out of the vine maples headed right for them; it was sliding over the top of other logs above them, and gaining momentum. They were standing no more than six-feet apart, as this tree passed between them. The top, embedded it's self in the brushy terrain, and the butt-end came to a stop between their feet. George said: "That tree came so close it touched my rain coat. I remember that as if it happened only yesterday."

Harold Fields (left) and "Shorty" Fenner (right) watch a large old growth log in process of being stored at the landing at Glenbrook for Fenner & Shroyer Logging Company.

We had a crew in the Cascades, up at Rhododendron, near Mt. Hood, where we were logging Douglas Fir for Pope and Talbot. The crew stayed in cabins and batched there during the week since it was too far to drive every day.

Carl Springer was driving truck for me and I was running cat, helping pull trucks out on those steep roads. After supper one night, I was noticing a nice watch Carl had just bought. He said: 'If anyone takes that, they will have to wrestle me for it.' That wasn't the thing for Carl to say to me. It wasn't long before I had him down in a toe lock and he was begging me to take that watch. I was fifty-five at the time, Carl was fifteen years younger. I was always going to give the watch back to him someday but Carl died before I got that done.

George and his crew at Rhododendron were missed by the terrible 1962 windstorm. The Columbus Day Storm was of the intensity you see in the Oregon Coast Range about once every century. The winds reaching gales over one-hundred miles-per-hour blew down thousands of trees. It happened on a Friday afternoon. Trees fell across logging roads trapping loggers in the woods. Some of them had to walk out and it took several days to reopen the roads. To log the fallen trees throughout the mountains and clear the streams and creeks of the debris took two to three years.

Rex had lots of blowdowns from the Columbus Day Storm that he needed out. He gave me a job getting those out so this gave me a chance to start a second crew working. I had only been back from Canada for a few years and was still in debt. I was sure glad to get a contract with Rex. Logging this was especially dangerous because it was on steep ground. When the butt was bucked, a root-wad or stump was left; these were real dangerous. A root-wad nearly killed Gail's son, Jim Shroyer. He was working for Hoyt Logging Company when one of those ran right over the top of him and covered him up. They thought sure he was dead. He was lucky there just happened to be a little place in there so he wasn't crushed, and enough air to barely keep him alive. They hooked the yarder on that to get it off him, took lots of skill to do that, he was real lucky he wasn't killed.

Those root-wads don't roll like logs; they are bigger on one end than the other. They are real dangerous on hill sides; I've seen guys try to out guess them. Those darn things watch a guy to see where he is going, then they turn right around and chase him. One chased Old Bill and I thought I was going to die laughing. I was watching him try to change directions to stay out of the way, that was the funniest thing, but it sure was dangerous.

Whether George was building roads, blasting, or grading he would always be looking for some fun. He was always playing pranks on someone or stirring up

Bud Carey operating a Hesselman yarder at Glenbrook west of Alpine on Prairie Mountain for Fenner & Shroyer Logging Company.

Diesel powered BU-50 loading donkey at Glenbrook. Merle Gee on left. Harold Fields, right.

some excitement whenever he could. They were working in an area where there were some especially large rocks when he noticed Herman Strom's Volkswagen parked nearby. George was running an Austin Western grader then.

> I thought it would be lots of fun to give Herman Strom one of those big rocks to drive around with. Since the trunk is in the front and the motor is in the back, I made a big hill of dirt right there beside the front end of his Volkswagen. I picked-up one of those three-hundred pound rocks with my grader, put it on that mound of dirt and rolled it right into the trunk of his Volkswagen. Closed the trunk and scooped the dirt away so he wouldn't notice anything different. Herman ran his Volkswagen for several weeks without knowing what was wrong. Herman told his son Gary: 'This thing acts awfully sluggish but I don't know what's wrong with it.' One day Gary drove it somewhere, he noticed that every time he hit a little dip, the front end would come down. He got out to take a look in the trunk, and he found that great big rock in there; it was nearly the size of the trunk. Herman knew right away how that rock got in there.
>
> I'd be up in the woods real often checking on my logging operations and whenever I'd see a lunch box laying around I'd spike it to a stump. In the hot summers before trucks were air conditioned, the truckers would leave their lunches near creeks or in shady areas. I'd look for places where they'd stop to hide their lunches. They'd use culverts pretty often, it would be real cool in there. I'd try to get to those before lunch time. Never liked the sandwiches too well so I'd leave them and take their cakes or pies. Always left the wrappers to remind them of what I'd eaten though. Sometimes I'd add bark or ferns in their sandwiches. They got to liking that stuff real well.
>
> One day while I was up at the logging site I noticed that Monte Shriber made him a chair from a log. He took the saw and just cut him out a nice seat and back. He worked as a chaser on the landing and I didn't mind him settin' as long as he got his work done, you know. That day, I decided I'd have some fun with that guy. I took some jaggers from a broken cable and hammered them in the seat of that chair real good. He was so use to settin' down in there; he didn't even notice those broken wires. Boy! He gave out a yell and up he came, those jaggers were real sharp. I laughed and said; "Monte, you sure seem in a hurry to get back to work."

Georgie soon followed in George's footsteps and he became an asset to George's logging business. From the time he was four-years old he preferred graders, dump trucks and cats as childhood toys. He loved to go to the woods so George would take him along whenever he could. If he didn't, Georgie would catch a ride with the truck drivers. As a child, he watched the perfection of Oscar Mock as he welded, dreaming of the days when he would model him.

> I never learned to weld, but Georgie is a darn good welder by anyone's measure. He was no more than a kid when he built that cab for the HD cat with the triple drum. Old Bruce was getting frozen out, up there on Yew Creek. He could hardly get over that kid doin' that. Georgie started whistle punking when he was only sixteen, and he was soon setting chokers. Mama worried Georgie would get hurt, but I was determined to teach him how to work.

Shroyer Logging Company was known for trying things new. Georgie often helped design original pieces of equipment as well as adaptations to some of their older pieces. Georgie also excelled in repairing equipment. He helped George keep the machinery in good working condition. Their crews had minimum down time since they did nearly all repairs themselves – and in short order too!

> One year our total expense for break downs was only eleven-thousand dollars and that was logging forty-million-board-feet of timber. Our accountant just couldn't believe it. I learned early in life to make your breakdowns minimal.
>
> I had bought that D-8 from Rex to start my own logging business. The track had worn out. Georgie went along to watch, he was just a little kid then. We worked 'til two o'clock on Saturday getting out logs, then Ernie and I tore the track off that D-8 and put new rollers on it. We were ready to log again on Monday morning. Then the next weekend, we fixed the other side. Another guy had the same kind of cat with the same problem. He sent his down to Pape' Brothers in Eugene and it cost him several thousand and two-weeks down-time.

Georgie became a partner in the Shroyer Logging Company in 1965. He and Annie Wischnofske married in 1967. By this time George was nearly out of debt and their financial prospects were more favorable. George had always been a working boss and now there was a team. George was beginning to feel the results of the many physical injuries he'd had over the years. Georgie recalls watching his father put on his climbing belt and spurs for the last time in the Siletz Gorge when George was sixty-years old.

> We moved our own logging equipment from side to side. Our first steel tower was a Berger Tower with a triple drum. Rigging up was easier than it had been in the past as the steel tower was rigged while it was down on its side, then raised. Occasionally, Georgie would have to climb a tail tree when they were logging out too far to get a lift.

When Georgie and I started out as partners, I got to do what I liked best. I did even more running the cat after that. I liked building a road better than anything because you could see what you got done. How would you like to work in a mill pullin' green chain all your life? Probably built over two-hundred miles of logging roads, graded, put in bridges and culverts. I also built another two-hundred miles of landings, spur roads and turnouts. While building roads, we would remove stumps and solid rock, this would all have to be blasted.

Blasting could be real dangerous, you had to follow some rules. I knew a guy who put his pipe into the same shirt pocket with his blasting caps – that killed him instantly. So, I always tried to be real careful with blasting caps. Once I had one that didn't go off. I noticed this guy watching me real close, he must have figured it hadn't gone off. Blasting always reminded me of Buck blasting that gas barrel flat. Ha! Ha!

I've built lots and lots of roads by myself with nobody around. I never felt I was in danger. I'd always cut out a flat break on a steep hillside. All these trees and limbs up above you want to come down when you are building a road on a steep hillside. Safety depends on the operator knowing what he can do with his cat. A few times trees fell right where I'd been settin'. I've had those things come down right across the seat with me out the other side. You've got to be real fast gettin' out of there, you know. Whenever I was working on a steep hillside I'd always try to kick out anything that might fall, before I got by it. Rex never let anyone build road by himself, but me. He told everybody: 'Anything George says is dangerous—nobody better try it'! He also said: 'If everybody says it's dangerous—it's a sure thing for George'! I've had a lot of luck! Joyce and Lucy think that I was given more than one 'Guardian Angel.' During my lifetime I've had over a hundred close calls, and some of them plenty close, I'll tell you!

Willamette Valley Lumber Company bought a large timber holding in the Siletz Gorge from Georgia Pacific. George and Georgie were logging there beginning in 1965 and working there until the end of 1969; logging over one-hundred-million board-feet of timber from that area. It was a difficult area to log since it was steep terrain on one side and bordered by the Siletz River on the other side.

We rebuilt the main road in there and Willamette Valley Lumber Company furnished crushed rock from a quarry they had. I built all the side roads and landings. We had a rock pit up Blue Jay Creek, we used that to make gravel. Bill, my brother-in-law "shot rock" and worked on the roads. First pit we shot we lost every bit of that rock, and it scattered over a-quarter-mile, we never got a load out of that. We bought our powder from Timber Supply in Philomath and they

sent a powder man down there. He told us what to do, there's quite a knack in shooting rock. When you know what you're doing you can shoot a large rock pit in one explosion and have it all heaped up in the middle when you're through. I don't think we lost a wheel-barrow full-of-rock after that!

We bought a new Berger yarder in 1966 to use up there at the gorge. I was pulling that yarder around Sunshine Ridge with my D-7E cat. We were moving it when the outer part of the road slid off. The back end of the yarder started to slide toward this steep canyon. The tongue twisted and broke loose from the cat, this was the only thing that kept the cat from going down the canyon with the yarder, that sure was a close call.

Shale rock had filled in along the hillside where we tried to take the yarder out. We should have just cleaned it out with the dozer. I didn't want to do that, I thought we could get through there without dozing it. When I felt it slide, I figured on getting to this big tree. I thought I could get my cat behind it, I didn't make it. If the dollie had broken loose first instead of the tongue fastened to my cat, then the tube would have come down on the cat, and jerked the cat down. If it had I wouldn't be here to tell you about it. The yarder stopped about two-hundred-fifty-feet down, the dollie went all the way to the Siletz River fifteen-hundred-feet down. It was the funniest thing you ever saw. It would bounce on its eight wheels, go way up in the air, come down again and bounce up. It kept doing this 'til it landed right-side-up on all eight wheels, right there in the middle of the river about one-hundred feet from two guys who were fishing. I can tell you they didn't waste any time gettin' out of there, they got in their pickup and away they went.

A couple of guys from Willamette Valley Lumber Company came walking up to me right after it happened. One of them asked; 'George is that your yarder down there'? I told them it was. That guy said; 'Then what are you laughin' for? Good Lord! If that were mine I'd be crying.' I said 'Well, if you think crying will help anything; if that will bring it back, then I will.'

While I was still at the Siletz Gorge, I had another real close call. Georgia Pacific had a steep piece they wanted me to log. There was a trail in there but I had to build a good road around this hillside. It was too steep to make a road through this one-hundred-foot draw so I was starting to fill it in, I wanted to raise it up about ten feet. I ran my cat up on the bank above this old road and I started to put the road in so I could kick dirt down there. Doug Isom was out in front and I told him to watch for me in case I started to slide off there. On a cat you can't see when it starts sliding sideways. I seen his eyes were getting 'great big'! I still had a footing, so I headed for this big tree hoping I

could make it but I never got to it. My cat went down twenty feet, right over on its side, and stopped right there on the old road. I moved pretty fast on the dozer's arm and jumped to the bank. It was too steep and I couldn't stay there. I fell right between the tracks underneath the cat. The tracks were going 'round and round.' After a minute, the oil pressure dropped and the tracks quit running. It never hurt anything. I had to get another cat in there to pull my cat out. I always figured out what I could do if something went wrong, before I ever got on a cat. When I was about to do something pretty dangerous, I always had an escape route planned out, you know. This helped me out of many dangerous situations.

The Fenner and Shroyer partnership ended with Eugene "Shorty" Fenner's untimely death in 1969. George bought Mrs. Fenner's share of the partnership, dissolved it and Shroyer Logging took over the contract with Pope and Talbot.

Charlie Hall, George Shroyers son-in-law, running an International cat loading a 1-log-load off a rollway onto a 1953 Autocar truck. Charlie Hall was sub-contracting with the Fenner & Shroyer partnership while George was in Canada.

CHAPTER 23

LOGGING DURING THE SEVENTIES

Changes commencing in the sixties developed into a distinct change in worker's attitudes by the seventies. This generation of youth became known as the "Hippie Generation." Their lives now reflected the disintegration of their moral fiber, the dramatic ramifications of a "war gone wrong" played its self out after the Vietnam War. The younger generation no longer looked with pride on the American way; they no longer felt that America could do no wrong. They had friends who had experienced a most nightmarish war. Many young and old alike felt that Americans did not belong in Vietnam.

Younger loggers, no longer endowed with the "American Spirit" as were the youth of the first half of the Twentieth Century, did not experience the camaraderie enjoyed by the young men of the earlier generations. No longer did the young loggers wrestle and scuffle – now they slept in the crummy on the way to and from work. Tricks with lunch pails and water jugs rarely took place and the young loggers were not as conscientious about coming to work or being on time. These young men of the Seventies were not "light hearted" as young men had been in the earlier days. They were now drinking more and taking drugs and this mixture with logging work was dangerous.

Mechanization had taken place and the machinery was now easier to operate. The size of the equipment had diminished; the chokers were a third the weight of the earlier ones. The remaining trees were half the size of the old giants that were previously logged. The loggers didn't have to work as physically hard as they had earlier.

> **One day Bud Carey, the guy we called "Blackie," was telling me about a bunch of young guys who were complaining about the strength needed for handling five-eighths-inch cable. He told them; 'When I was a young guy your age, in the early days of logging—**
>
> **Trees were made of Wood, and Men were made of Steel.**
>
> **- - - - Now the - - - -**
>
> **Men are made of Wood, and Trees are made of Steel.'**

Today, spar trees are made of steel. Men don't have to climb these steel trees. That has all changed. Also, even if men were still climbing; only a few trees, large enough to be spar-trees, remain.

> **One fall, Pope and Talbot wanted me to log ten-million-board-feet more than I was already logging for them. I didn't want anything more; had all I could do. I thought, if I didn't turn in a bid they would think I wasn't interested. So, I turned in a price of ten-dollars more per-thousand board-feet. That was quite a bit higher than I'd been getting, I figured they would get somebody else to log it if I did that. I**

George Shroyer running his cat against the 'big log" old growth Douglas fir (10 ft. 6 in.) on September 22, 1976

felt pretty smart about that; I didn't want to do it anyway. One of the guys from Pope and Talbot came out to talk about it. He said: 'George, you just don't want that job, do you'? I knew him pretty well, so I said: 'No. I don't'! He said: 'We've talked it over at the head office; we're just going to give it to you at your price anyway.' So, there I was, stuck with ten-million-board-feet more and only two months to get it done in. I had to find more equipment and help. The only way they gave it to me at that price was because they thought it would be real funny to see me rushing around to get that done, I'll never forget that. It was quite a job, only thing is; it paid good.

Rex Clemens sold his timber holdings in the Alsea and Harlan areas in 1972 to Willamette Industries, previously Willamette Valley Lumber Company. George and Georgie began contracting additional logging and road building, negotiating prices for each new job with Willamette Industries.

Georgie and I were now running three high-lead-logging sides and one cat-logging side. We kept all our equipment repaired and helped move it from site-to-site. Bruce Hamm, Ernie Solie, John Eagleson and others ran the cats, they were real good road builders and cat loggers. Bob Eggers ran the shovel loading out logs on the cat side sometimes, other times he loaded out logs from the right-of-way where we were road building. Charlie, my son-in-law, loaded logs off the right-of-way for thirteen years for us; he also had the contract to haul our logs from all our logging sides with Hall Trucking. We did most of the road building for Willamette Industries in the Alsea area. I laid out and built roads when I had time. We hired Bud Carey, Bruce Hamm, Doug Isom, Ernie Solie and others to be 'woods bosses.' A short time later, we hired Bud Carey as Superintendent of Shroyer Logging.

All of these men worked for me for thirty or forty years except Doug Isom, he was killed at Fall Creek. He was a good natured, hard working, red-head and he did a good job for us. He always wanted to get things done. I remember the time Doug was working on the steel tower and he had his head stuck up in that tower. When I see something like that, I just can't help it -- I grabbed a sledge hammer and hit that tower as hard as I could. Doug said that his ears rang for several days after that, he should have killed me!

We were working for Willamette Industries a couple of miles north-west of Alsea on Follet Road. It was a bad show and wouldn't have made alot of money unless I came up with a cheap way to log it so we did what we called "double-yo-yo yarding."

George Shroyer with a D-7E cat and a Washington 78 swing yarder in rear. Scene is on Follet Road where George did "double- yo-yo-logging" in a very steep area.

It was real steep and we couldn't put much of a road in there. There weren't enough stumps to drum behind either. Our yarder wasn't big enough to take some of these big hooters out. So, I put a big block for the cable on a large stump at the landing. Then I built a pretty straight road down the ridge, it was wider in the center so two cats could pass. We hooked a cable to the back of the cat going downhill and the other end was hooked to the front of the loaded cat coming uphill. This way, you know, the weight of the cat going downhill helped pull the loaded cat coming up the hill. When the cats came to the end of the road I'd built; the chaser would unhook their line and move it to the other end; depending on whether it was headed uphill or downhill. We had a chaser on each end to unhook these. Willamette Industries thought it was a pretty good deal; we made some money out of that.

I remember some stories, one was about John Shackelton and the others were about Mike Mintonye, they worked for me back then. John Shackelton and Jim Steeprow ran the two D-7's and Bob Eggers was loading with the 98-Link-Belt loader. One day, I was up there cleaning the landing with my cat. Mike Mintonye, our company truck driver, was looking out the back window watching Bob load Mike's Peterbuilt truck. I drove my cat over until the front idler was up against the front wheel of his truck and the corner bend of the blade was right over the center of the hood. We were both lookin' at each other; but we weren't, you know. I brought the blade of my cat to within a quarter inch from the hood of the truck and it was clearing the fender. I saw it go down just a little bit and couldn't imagine why, then I saw it pop back up. I forgot that little reflector on his fender, we all had a good laugh over that.

John Shackelton first worked in the garden for Mama. He was wanting to get out of the garden and work in the loggin'. He started working for Georgie and I, he worked for us during the seventies and eighties. One day, up at Rack's Creek, we were building roads, I was running the cat and John was helping put in the culvert. I made a ditch for it and John was doing the hard work. We'd started back-filling it, I got a scoop of dirt off the bank with my cat. A little alder, about twelve-inches on the butt and about thirty-to-forty-feet tall came out with it and I pushed it down to where John was filling in. It had a bow in it and as it straightened out John could see the top of it would reach him, he dove down into the mud. I laughed so hard I had to stop the cat. John was all covered with mud. He crawled up on my cat and walked right across my feet and down the other side with his cork shoes. All the time, I was laughin' so hard I thought I'd die.

Burger yarder with a Scookum-Madill Tower bringing in a turn of logs to the landing.

Mike Mintonye, our truck driver then, only got two tickets for overloads in the sixteen years he drove for me. He was real good at 'settin' out.' Sometimes, when a trucker was overloaded he'd wait until the weigh master left for the night. Mike told me, one time during his trucking days. the weigh-master stayed up all night waiting for him. Someone had seen him as 'settin' out' and reported him.

There were some highway weigh-stations beginning in the mid-forties, but we didn't get one in Philomath until about 1955. Even then, they had some portable scales. These were carried around in the back of a pickup truck. These portable scales only weighed about a-hundred-fifty-pounds apiece. They were built like a bathroom scale, but they had a little ramp-like place for you to drive onto, there was one for each side. Whenever they would see you on the road, you'd have to pull over; they'd get those scales out and check to see if you were overloaded, if you were, they'd fine you.

Whenever you had a one-log load, you'd almost always be overloaded. You couldn't take a log off to make it lighter, in a case like that.

If you had three logs on, they'd make you take one off before they'd let you go on your way. You'd have to leave your truck there and call in another truck to take the other log. That would make some truckers pretty mad. It was interesting to hear all the different excuses the trucker's used when they were overloaded. Probably, the most honest excuse was: 'I didn't know the weigh master was going to be open.' Other excuses would be: 'Our scales aren't working right.' 'It's not the truck I'm use to driving.' 'We've just moved to a new landing.'

Willamette Industries had some good supervisors that looked after their interests and oversaw the logging of their timber. Those guys treated us real well. Norm Marsh, Leroy Volz and Terry Aman were some of them. I always told my crew: It's OK to joke with each other, but leave those Willamette Industries guys alone. It could cost us our job.'

I remember Willamette Industries would send Willie Hallowell, another supervisor, out to help us burn our slash in the fall after the first rain. This guy hated to walk fast, but with a fire behind him; I tell you he would move. I'd get down below him and start the whole hill side afire. Boy! He'd have to scatter to get out of there. Burning was worth it just to see Willie move. Having fun was important but keeping the job was always top priority. The job came first.

George had an enormous amount of blasting experience, he'd built many miles of roads using powder to help remove stumps and root-wads while clearing difficult areas of rock and debris. After buying their six-hundred-forty-acre ranch at Missouri Bend, he blasted out hundreds of stumps to clear it. George also used dynamite to clear three-hundred acres at the Detroit Dam.

We were working below Alsea, near the Bethel Vernon Place. My crew was getting ready to blast out a one-thousand-foot ditch in a swampy area. The BLM was supervising it and some of their guys and an employee from Timber Supply were there. We were using several boxes of ditching powder to create a domino effect. It would only take a few seconds to move from one end of the planned ditch to the other and we were all ready to set it off. We'd had previous experience with powder so I said: 'You guys better pick yourselves out a tree and get behind it.' They said: 'We'll be fine, the book says that stuff won't go far.' I said: 'OK, but you won't mind if I do will you'? So, I headed for a large cluster of maple trees that had lots of smaller bushy trees around them. I wasn't there more than a few seconds before they could see it was going to get them. They all jumped right in there on top of me. The bushes and trees, we were behind, were just covered with that mud. The pickups were standing pretty close and they were all covered with mud too. One of the guys even got a broken windshield. You never know what's going to happen with ditching powder, it can work awfully good in wet places.

While we were working up at the Siletz Gorge one time I helped clean out a septic tank. Vern Bright lived up in Siletz and he was one of the guys that ran the square-tower Berger yarder for us, sometimes he'd work over the weekend. This one weekend, he said his wife wanted him to stay home to clean out the septic tank. He asked me if I thought one stick of dynamite would do the job, and I told him it might take two or three sticks. It cleaned it out all right, everything within a block was covered, and it smelled awful. It took-off part of the garage roof, and broke the windows. It was all over the rest of the roof, and their house. Two boys across the street were working on their car, they were covered with it. They thought that was pretty funny.

When he came back on Monday, Old Vern started laughing so much he could hardly run the yarder. Guess all his neighbors called him up and wanted to know if he could use another job. They said their septic tanks needed cleaning too. Vern was laughin' so hard he just couldn't work. So, the whole crew had to shut down for awhile 'til he could run that yarder again.

Gary Strom worked for me falling timber, he was really a good man. Only problem was; he just couldn't help laughing even at something that was serious. One summer back there in the seventies, we had a contract with Willamette Industries falling timber up there on the watershed. One day, some supervisors from Willamette Industries were up there inspecting the job site, we were standing there talking and watching Gary Strom falling those big trees along the creek. Gary wasn't having any trouble getting those trees to fall away from the creek like they were supposed to. While we were standing there talking, those guys were telling me how important it was to keep those trees from falling in the creek. It wasn't like it is today, when they want you to leave a few big trees in the creek.

There was this one big tree that was leaning way out over the creek; Gary was cutting away on it and he'd almost finished it. We were watching this tree to make sure it didn't fall that direction. This old tree just kinda stood there swaying, it didn't fall right down like they usually do. All at once, a big gust of wind came down the hillside and blew that big old tree right into the creek and there were all those Willamette guys watching that. Well, that kinda struck Gary as funny. That big old tree falling into the creek with those guys watching. He couldn't help himself, he just started laughin' and he couldn't hardly stop. Well, those Willamette guys didn't think that was funny at all, they got real mad about that. Those guys thought Gary didn't give a damn. The Willamette boss looked right at me and he said; 'Get rid of that guy, I don't want him on the job any more.' So, I had to let old Gary go, just because he couldn't help laughing. I recommended him for another job, and they were real happy with him. As it worked out, Gary finally ended up working for Willamette Industries again.

We spent alot of time cleaning out those creeks—a guy could see that was the wrong thing to do. Cleaning out the creeks, let the water run too fast when it rained alot. It'd flood, cause there wasn't anything to slow the creek down. I'd go back a few months later and there would be these big wide muddy places. Lots of times, there wouldn't even be a creek bed there any more. Just a place along the stream, sometimes twenty feet or more wide, with about six inches of water running over it. Later, they started sending guys back in there to fall some trees in those places. A forester telephoned Ermie Walter awhile back, he was wanting to see if Ermie would volunteer to fall some trees in that creek for them. I told Ermie they should get hold of Old Gary Strom. He'd sure get a kick out of that, you know.

Jack Bowen, one of George's fallers, ran the crews that fell some very large trees in the Alsea area. The tallest tree was logged for Pope and Talbot, it was fell near Alsea Falls, in 1972. It was two-hundred-seventy-four-feet tall, nearly as tall as some of the redwoods. Later in 1976, Jack Bowen along with Leroy Russell fell one of the Alsea area's largest trees near Skunk Creek for Willamette Industries. This Old Growth Douglas Fir had a ten-foot-six-inch diameter and its age was estimated as four-hundred-twenty-five years. The big fir was more than one-hundred-eighty-feet tall; and contained more than forty-two-thousand-board-feet. Jack Bowen and Leroy Russell used a "spring

One of Alsea Area's Largest Trees. This eighteen foot butt-cut old growth Douglas fir with a diameter of 10 ft. 6 in. contained 9,000 board-feet. It was taken out of Skunk Creek along the Alsea-Winney Road on September 22, 1976. Pictured left-to-right: Dan Winney, Le Roy Russell, Jack Bowen, George Shroyer. This prize log was hauled in parades for the Albany Timber Festival and other parades.

board" while cutting this giant, they cut it into seven sections. The hooker for this was Dennis Andrews of Blodgett; Sam St. Clair and crew did the yarding; Bud Carey, Woods-Superintendent for Shroyer Logging ran the shovel that loaded this giant log.

> We were all there, Georgie and I and our crew, even Charlie and Joyce came up there that day during the yarding. We had them stop the logging for half-an-hour so we could all get pictures of it. Sam St. Clair and Joyce sent the information and photo of that huge tree to the newspapers. Willamette Industries kept that log in their Dallas yard that winter. The next summer, Wimers hauled it in the "Albany Timber Carnival" and other parades.

Special hauling permits were required, due to its enormous size. It was hauled on a low-boy used for hauling large pieces of equipment. It was hauled to Willamette Industries, Dallas yard, by Jim Goodwin of Philomath on the Shroyer Company's low-boy. It was one of the largest trees ever logged in the Alsea area. (See front cover of this book)

George and Helen Shroyer on their 50th wedding anniversary.

George and Helen wanted to invest in another ranch and began searching for a ranch in Eastern Oregon. They heard there was a ranch for sale over near Vale, the "Harold Jenkins Ranch." George and Helen soon visited the ranch. This place had three houses, an upper house, a middle house and a lower house; it was flat land and rolling hills with creeks, used for irrigating about 1,000 acres of fields. The first time they saw it they told the Realtor; "We'll buy it"! Soon an adjoining ranch, the "Fenton Place" was for sale; and they bought that as well. Together these two ranches contained 9,000 deeded acres of land and 41,000 acres of leased grazing lands.

George and Helen soon incorporated their three ranches, the Salmon River Ranch, the Vale Ranch, and the Alsea Ranch into Shroyer Ranches Inc. Helen moved the Gellatly Canyon home Mr. Hunter, her father, built for them, to an oak knoll near the "Home Place." Their youngest son, Rex, and his wife Holley, moved into the home where they are rearing their children. Rex soon became manager of Shroyer Ranches Inc.

We have several smaller farms around Philomath that Mama and I own. Mama worked on all of them raising cattle, she did lots of moving irrigation pipes and spraying weeds. Georgie and I helped by putting in the hay, doing the farming and repairing the farm equipment. I always joked about Willamette Industries shutting down around the 4th of July. I told everyone; "They did that so I could get my hay in!"

When we got married, Mama and I thought if we could own forty acres, we'd be happy. Now we have forty-thousand acres. It's funny how a person's idea changes.

Mama and I celebrated our "Fiftieth Wedding Anniversary" on December 24th, 1976. During those years I made a fortune, went broke, got out of debt, and made a fortune again. Mama and I have been through the 'Great Depression,' and several unexpected family deaths. Those were awfuly hard to take. We lost a baby and a young son-in-law, then three grand-daughters. First Lucy and Jimmy lost two daughters while they were very young. Then, Georgie and Annie lost Holly, their little girl, in an accident when she was only seven years old.

I've seen lots of tragedy in the woods, when we lost Buck, it was like losing a brother. I've seen about a dozen others killed in logging accidents too. I've been through good times and bad times but mostly good.

I've been real lucky to have had great people in my life who influenced me. First was my dad, Oliver Shroyer, then Mac McCoy who owned the mill at Kings Valley, later Rex Clemens and T. J. Starker. The last time I visited with Starker was at our anniversary, he was walking with a cane that day, Mama and I were really glad to see him. I owe my success in many ways to all four of them.

CHAPTER 24

LIFE DURING THE EIGHTIES & NINETIES

In 1980 I had an accident with my knee and I never worked out in the woods much after that, I was seventy-three years old then. Georgie took over all the logging operations for me. After that, I spent my time repairing logging and farming equipment in the shop, farming the 'Home Place,' and helping Mama with the ranches. On my eightieth birthday, Shroyer Logging crew gave me a party at the shop. That was in 1987. Georgie and Annie brought over a great big cake and we had soft drinks. Not many loggers have a party like that. Mama and I never liked the drinking most loggers are known for.

The older I get, the more I realize that money spent on trips is money well spent. My Dad always said: 'Traveling is part of your education.' During the seventies Mama and I traveled to Europe twice on an American Charlois tour. We visited England, France, Italy, Holland, Germany and Switzerland. Our group visited several ranches to see what European Ranch practices were. It made me realize that land was always going to be a good investment, land over there was real high. They're not going to make any more of it, you know. So, I figured land would increase in value here, over time. The land I've bought has gone up more than money left in a bank would have.

When Rex, my youngest son, was managing the ranches he joined the American Angus Association. This organization sponsored a tour to Australia and New Zealand in 1987. Mama and I wanted to go, so we took Rex and Holley, and Lucy and John Leach, Lucy's new husband. We also decided to invite Mama's brother Bill and his new wife Emma, to come along with us. Doris, Bill's first wife, had died in the early seventies. We were going to visit ranches in the Australian Out-Back Country. I was already eighty-years old and beginning to walk slower as I was pretty stiff, and my balance wasn't too good. We had a real good time though.

Wherever I went, I always had fun. There was a woman named Virginia that went along on the tour. She had a great big wheat ranch up on the Columbia River by (the town of) Boardman. She had already sold part of her wheat that fall at a good price. She was watching the market real close, trying to get a better price on the rest of it. Virginia was pretty good natured, so one morning I said: 'Did you hear what the wheat market did this morning'? I pointed my thumbs downward. She got real excited, and she took out of there to get to a phone. She came back with a silly grin on her face and gave me heck for lying to her. Whenever I saw her after that, I'd point my thumbs either up or down. Everybody got a kick out of that; it was

Fellows at the shop celebrated George Shroyer's 80th birthday. George is 7th from left. Georgie stands at right.

pretty important to her. She still had several thousand bushels of that wheat left to sell.

The American Angus Association sponsored another trip in 1989, this one was a trip to South America. My balance was getting worse and I'd just taken a real fall in my shop that year. Walking was getting to be pretty hard for me, and Mama's knee-joint was wearing out and she was having lots of trouble walking too. Mama and I went to see our orthopedic doctor, in Corvallis, and we asked him if he thought we could go on this trip. He encouraged us to go – as long as some of the family went along. I'm real glad Mama was so determined to go on that trip. Since then, we haven't been able to do much traveling.

That time we took Joyce and Charlie, Lucy and John came along again, so did Bill and Emma. They were always a lot of fun to have along. It's a good thing the girls and their husbands were in pretty good shape because Mama and I both had to use a wheel chair to get in and out of the airports and through customs. It would have been too tiring if we'd been standing in line all that time, that would have been too much. Being pushed around in a wheel chair was real hard for me to get use to. I was always the one that took care of others before that. I never thought I'd get to the place where I couldn't walk. When you're healthy all your life you never think of things like that.

Once we got to South America I could use a crutch on my right side to help me balance a little better. Then I could hobble along at a fairly good pace for an eighty-two-year-old man. Mama and I and Bill were all getting pretty old by then, we all had trouble getting in and out the tour bus. The kids always let us set up front. We'd get out and go on the walking tour if it was something real important, otherwise we'd set in the bus and just look out the windows. The kids always brought us something, a toy or some candy. One time John brought me an imitation snake. It was about twenty inches long and it was made out of diamond shaped pieces that moved. If you squeezed it behind its head, its mouth would open and its tongue would jump out at you. I would set there in the bus with this snake around my arm and every one got a kick out of it except Robare, a Columbian. He turned white as a sheet of paper the first time he saw it, and after that, he called me Mr. Snake. I decided to leave that snake behind but, I couldn't even give it away. Nobody wanted it, and I didn't want to be responsible for somebody having a heart attack.

The South American Angus Tour took in Chile, Argentina, Brazil and Paraguay. There were lots of opportunities to buy ranches in those areas, the land over there was very inexpensive. Sure wished Mama and I were young again, we could have bought a Cattle Ranch over there, Mama would have liked that.

When the Shroyer's returned from South America, George's condition worsened. He began having more trouble walking and he was loosing the strength in his right arm. The neurological tests showed there was severe pressure on George's spinal cord. This condition is known as cervical stenosis. George was told that it was a progressive condition and it would lead to a complete paralysis from the neck down. The Neurosurgeons consulted were not at all optimistic about a successful outcome after surgery. George, determined to continue to work, helping with the haying that summer.

That fall, with his condition worsening, the family became insistent that they find a skillful, confident surgeon. Dr. Bruce Byram, their family physician, referred them to several potential surgeons. This resulted in the family finding a surgical team, specializing in neck surgery and consisting of an orthopedic surgeon and a neurosurgeon. This team of surgeons worked at the Hospital in Salem, Oregon. They agreed to operate on George's neck. The neurosurgeon told Joyce: ‘We're facing five major problems with your father's condition, I want you to tell the family but I don't want you to tell your father, I don't

want a frightened patient.' The only remaining job was setting a date for George's neck surgery. The year was nearly over and the earliest date possible was near Christmas.

Early one snowy December day in 1990 the Shroyer family took George to the Salem General Hospital. His surgery was scheduled for noon. Their prayers were answered, George made it through the surgery. The two surgeons were relieved to find, that in only a week, George was ready to go home. He'd had only a short stay at the hospital during the Christmas Holidays. Helen welcomed him home by tying a big yellow ribbon around one of her prized oak trees near the entrance to the 'Home Place.'

George convalesced in a hospital bed they'd moved into the dining room. He arrived at home in a special vest having a metal halo permanently attached to it. They sure gave him a bad time about having a halo that Christmas before he'd even made it to heaven. George said: 'Never mind me, I'm just trainin' to be an angel.' Both Joyce and Lucy moved in with their parents to help care for their father that winter.

> **That halo was quite the piece of equipment. The vest had metal chest and back plates built into it and there were four metal rods fastened to the plates in the vest. At the end of these rods was a metal ring; into that ring, four metal screws were inserted into my skull. This made it so I couldn't move my neck at all. I think they gave me this halo to wear thinking it would teach me a little patience.**
>
> **It wasn't too bad, the government even did their part to entertain me. They started the Persian Gulf War on January 16, 1991. I laid there day after day watching them report on that darn war. My head was suspended in that halo the whole darn time. A couple of months later, when they took that thing off it left four holes in my scalp from those pins. When I saw those holes in my head, I told Joyce and Lucy: 'I think, instead of growing wings, maybe I'm startin' to grow horns.'**
>
> **By early spring I was beginning to regain the use of my legs and arms. I could get around some by using a crutch or arm canes. I worked real hard on getting strong again since it wouldn't be long before haying would start, and I wanted to be ready to do my job haying. When it came time to run the bailer I was doing real good. I ran the bailer for a couple of summers after that. Haying was my job. That was my way of helping Mama with her cattle. I was real determined to do my part of the work.**

During these years, Helen began asking George to build her a new bridge over to her flower gardens across the river. George planned and designed the bridge. He'd certainly gained a significant amount of experience during his lifetime, building his first bridge while only twenty-four-years old. This new bridge was to be seventy-five-feet long and about four-feet wide. Its frame was

made of structural steel with cross bars about every four feet. It even had a trailer hitch welded onto the front end of it. This bridge had been welded together by his neighbor according to George's specifications. It was to have a safety-railing, to protect Helen during her many trips over to the gardens on the other side.

> Once the steel frame was finished, we put a flat-bed trailer under the frame, this held up the center and gave it two sets of wheels. Then, we hooked the pickup truck to the hitch, and drove across the fields on the opposite side of the creek. When we were across from the house we unloaded it and left it setting on posts on the ground. That bridge was going to be pretty difficult to get into place. It lay there a long time.
>
> Mama kept saying: 'George, when are you going to get that bridge in'? I kept telling her 'We'll get stuck, it's still too wet.' She was going into the hospital to have her knee operated on. I planned on doing it while she was in there. We were going to move some of her shrubs and plants out of the way. We were even going to use her prize oaks to hold the skyline while we set the bridge in place. Boy! She'd have had a fit, if she knew we were going to do that and I didn't want Mama gettin' all upset.
>
> Once Mama was safely in the hospital, we got real busy. We got those plants moved right away, then we went to work putting up those cables for the skyline. These went around the oaks pretty high up so we stood timbers upright around each oak to keep the lines from cutting into the bark. If I'd hurt those oaks that would have been sure death for me. We used several oak trees so the bridge's weight wouldn't break them. It took most of one day to get that cable strung across the creek. We ran a skyline across to the D-7 Cat with the high arch on it. I ran the drum on the cat and Georgie and Rex did all the work. When we got that skyline in place, we had two ten-inch pulleys on there, then we cabled the bridge to the skylines. We tied the haywire to the bridge and ran it across the creek and fastened it to the pickup. We pulled that bridge into place by running the pickup up the driveway.
>
> It looked just like a logging operation going on in Mama's flowers. I just laughed and laughed to think what she would say if she'd been there. I don't think she ever guessed why we waited until she was in the hospital. The 'Home Place' sure looked like a logging camp with all those lines tied to Mama's oak trees, it's a good thing she wasn't there to see all that.

It had been three years since George's surgery, and he had been out there during haying season each summer doing his work. His asthma was worse each year and he was having more trouble getting up into the tractor cab. The exertion of climbing in and out left him gasping for air. Sometimes, Rex or John would help him climb into the cab. Once in the air conditioned cab, after resting, he could breath well enough to run the tractor pulling the bailer that made the big round bails. The smell of fresh-cut hay and drying bales was something that George looked forward to each year. George was out in the fields as soon as possible. He always said: 'The weather will give you enough time to get the hay in if you will do it.' It was getting to be pretty difficult for him but he wasn't going to be a quitter. He was determined to help them. He'd tough it out.

When haying was over that last fall, the family decided to consult the two surgeons who had operated on George's neck about his worsening condition. George's mobility was declining and the family was told that the lumbar stenosis in his lower back would require surgery. He was now eighty-six-years old and had pulmonary disease and was at high-risk for surgery.

George was again, taken to Salem General Hospital, this time for an operation on his lower back. Five days after this surgery there were complications with blood clots forming in one of his legs traveling to his lungs. The clots in his lungs created a critical situation as he was already suffering from chronic obstructive lung disease. George's five-day-hospital stay lengthened into sixteen-days.

Once again, he was brought home to the family that loved him. This convalescence was not as long as the first one. There was no halo to deal with this time. By spring George was thinking about the coming haying season. He would soon be bailing hay again. By mid-spring, George went to see his surgeons for a final check-up. During this visit, one of the surgeons said: 'Just shows what a determined, tough-guy can do if you give him half-a-chance.'

George was out there bailing hay for two more summers after his surgery. He was eighty-eight-years old by then and this was the first summer this writer had known George. Once the guy's helped him up into the tractor he'd wave them off. He'd say: 'You go on and get some work done, I'm OK.' His only problem was, the only way he could get out of the tractor was to jump. That is just what he'd do. When he got through with a field, he'd jump out of the tractor. Everyone was worrying terribly about him being out there by himself, but George wouldn't hear of anyone staying there with him. That's the way it was, he wasn't going to quit doing his job. George was just finishing up the last of the haying that summer. When the writer arrived one day, there seemed to be quite a bit more buzzing around than usual. I asked George what was going on and he just grinned and said;

> Well, I had nearly finished bailing the Roberts' field on the back side close to the river. I was driving that closed-cab tractor with the bailer on behind me. There was a loud noise when the universal hit the draw bar -- I knew the universal joint had broken. That caused some sparks

to fly and a fire started in the field, the reel on the bailer took that fire right in there, you know. I knew I couldn't get out to put the fire out and John was on the other side of the creek raking hay. Just as John jumped into the creek headed right for me, I put that tractor into high gear and took off out of there. That fire was spreading fast through the field, and I didn't want to burn up in that rig. I was spreadin' that fire as I drove out of the field. Figured I'd just drive over to the shop where Rex was working and he could hose it down to get that fire put out. Boy! I'll tell you! I had that thing going about twenty-two miles an hour, and I wasn't even out of the field yet. I was heading for the gate as fast as I could go. Drove that rig right out of thcrc, onto the main highway. Everybody could see the flames, smoke was just pouring out so, you could hardly see. They sure got out of my way in a hurry! I was heading for the shop, about a quarter mile down the highway. Some guys were standing in their driveway as I drove past. They started runnin' down the highway, pointing, and yelling that I had a fire on behind. Heck! I knew I had a fire behind me, I wouldn't have been going that fast for nothin'. Rex was near the barn next to the shop, and he helped me by wetting that fire down. We had to dump the bail that had caught on fire into the creek, it wouldn't go out. Miller's Tree Service, across from the burning field, came over with a couple of fire fighting trucks. The Philomath Fire Department was out there too. Pret'near burned that field up. The kids told me I wasn't going back to haying anymore. They said they wouldn't mind loosing the hay, but they sure didn't want to lose me. I sure do miss haying. I guess I always liked to work, especially when I could be running a piece of equipment.

By the way—I just received the electric wheel chair I ordered the other day. It came on my ninetieth birthday. I'll have a piece of equipment to drive now. We're going to build me a ramp down to the bridge so I can go across the creek this summer. I told the kids; "Now, if someone will buy me a van I'll be out-of-here."

Georgie has been managing the Shroyer Logging Company since about 1980 and continues to log in the Alsea area for Willamette Industries. He maintains that Shroyer Logging Company has been lucky to have worked over thirty years for Willamette Industries – a top company in the forest industry. Furthermore, Shroyer Logging Company has always been fortunate in hiring and keeping a good crew.

Logging continues to change. There is more competition and regulation now and the margin between profit and loss is much more narrow than it was in the past. Today the smaller size trees require newer methods of logging and modernization of equipment.

Willamette Industries requires that Georgie attend seminars provided by them on safety in logging and in protecting the environment, as well as new logging practices and technology. This has earned Shroyer Logging Company certification as Professional Loggers, and prepares the company to compete in the 21st Century.

Bibliography

Georgano, G. N. and Carlo Dem. *Trucks, An Illustrated History 1896-1920.* Edita S.A. - Lausanne (France), 1978.

Geographic Names Information System (Oregon) U.S. Geological Survey. 1992.

Helbock, Richard W. *Oregon Post Offices 1847-1982.* La Posta.

Hurwitz, Howard L. *An Encyclopedic Dictionary of American History.* Washington Square Press. 1968.

McArthur, Lewis, L. *Oregon Geographic Names.* Oregon Historical Society. 1992.

McCulloch, Walter F. *Woods Words; A Comprehensive Dictionary of Logging Terms.* Oregon Historical Society. 1953.

Pierre, Joseph H. *When Timber Stood Tall.* Superior. 1979.

Webber, Bert. *Oregon Names How to Say Them and Where Are They Located*? Webb Research Group. 1995.

__________. *Silent Siege III; Japanese Attacks on North America in World War II, Ships Sunk, Air Raids, Bombs Dropped, Civilians Killed.* Webb Research Group. 1997.

Webber, Bert and Margie Webber. *This is Logging and Sawmilling.* Webb Research Group. 1996.

Working in Wood; Illustrated Manual of Tools, Methods. Putnam, 1980.

Index

Page numbers shown in **bold-*italic*** type are photographs